T0230500

Fundamentals of Nanomechanical Resonators

Silvan Schmid • Luis Guillermo Villanueva •
Michael Lee Roukes

Fundamentals of Nanomechanical Resonators

Second Edition

 Springer

Silvan Schmid (iD)
TU Wien
Vienna, Austria

Luis Guillermo Villanueva (iD)
École Polytechnique Fédérale de
Lausanne, Switzerland

Michael Lee Roukes (iD)
California Institute of Technology
Pasadena, CA, USA

ISBN 978-3-031-29630-7   ISBN 978-3-031-29628-4   (eBook)
https://doi.org/10.1007/978-3-031-29628-4

This Springer imprint is published by the registered company Springer Nature Switzerland AG
The registered company address is: Gewerbestrasse 11, 6330 Cham, Switzerland

Preface

Back in December 1959, future Nobel laureate Richard Feynman gave a visionary and now oft-quoted talk entitled "There's Plenty of Room at the Bottom." The occasion was an American Physical Society meeting at the California Institute of Technology. Although he did not intend it, Feynman's 7000 words were a defining moment in nanotechnology, long before anything "nano" appeared on the horizon. The breadth of Feynman's vision is staggering. In that lecture 42 years ago, he anticipated a spectrum of scientific and technical fields that are now well established, among them electron-beam and ion-beam fabrication, molecular beam epitaxy, nanoimprint lithography, atom-by-atom manipulation, quantum-effect electronics, spin electronics (also called spintronics), and microelectromechanical systems (MEMS) or, even smaller, nanoelectromechanical systems (NEMS).

It is the latter topic which has been the focus of our research careers: mechanical systems with at least one dimension below $1\,\mu m$ (10^{-6} meters) where the number of atoms departs from what is usually considered *macroscale* and enters into the *mesoscale*.

NEMS have been developed for a bit more than two decades now. Fabrication technology has evolved and improved enormously for these devices which allows for reproducibility and yield, two *sine qua non* conditions for an eventual commercialization and, with it, a direct impact on society. The two driving forces for NEMS research have been *metrology* and *fundamental science*.

Matter at this mesoscale is often awkward to explore. It contains too many atoms to be easily understood by straightforward application of quantum mechanics (although the fundamental laws still apply). Yet these systems are not so large as to be completely free of quantum effects; thus, they do not simply obey the classical physics governing the macroworld. It is precisely in this intermediate domain, the mesoworld, that unforeseen properties of collective systems might emerge. Indeed, many efforts have been invested into cooling mechanical resonators down to their *ground state*, which has been recently proven. In addition, interaction between NEMS and q-bits, operation within Coulomb blockade, etc. have been also studied over the years.

On the other hand, NEMS can also be used as extremely good sensors. After the first pioneering experiments on measuring the quanta of electrical conductance and the one of thermal conductance, NEMS have also been targeted to detect spins, radiation, temperature, mass, etc. The theoretical limits for the performance of NEMS-based sensors are outstanding. However, these devices have seldom made it to the market. Comparing to their *bigger brothers*, MEMS, which it took them around 35 years to start having a great impact in consumer market, we can learn several lessons to foster the applicability of NEMS:

- *Standardized fabrication*—Fabrication should be reproducible, with high yield and the fabrication process should be standard (or very similar) across many different foundries.
- *Education*—An educated and trained workforce is required to tackle the problems that arise within these systems. Unfortunately, almost no university in the world offers a course in these type of devices, covering both the electrical and mechanical aspects.

It is the latter point that this book aims to address: to be a textbook for a course for engineers, not going into the details of atomic-scale simulation and analysis but rather taking an approach such as "top-down," i.e., using macroscopic formulas to model the devices.

At this point, it is important to note that this book focused on NEMS, that is electrically transduced nanomechanical resonators, in contrast to cavity optomechanics, which fully transduces and controls nanomechanical resonators by optic techniques. However, in particular, the mechanical models introduced in this book are key to the understanding and optimization of nanomechanical resonators used in optomechanics. There are also natural overlaps between the fields in hybrid devices, as it is the case, e.g., in microwave-based cavity optomechanics, which is based on electrostatic transduction but uses optomechanical techniques.

Vienna, Austria   Silvan Schmid
Lausanne, Switzerland   Luis Guillermo Villanueva
Pasadena, CA, USA   Michael Lee Roukes

Contents

List of Symbols

A	Cross-section area
α	Thermal expansion coefficient
B	Magnetic field strength
β_n	Wavenumber of mode n
c	Damping force coefficient
c_E	Wave velocity inside a specific material
C	Capacitance
c_p	Specific heat capacity
C_{th}	Heat capacity
c_0	Wave velocity
d	Diameter
D_E	Flexural rigidity of slender beam
D_P	Flexural rigidity of plate
δ	Phase difference
E	Young's modulus
E'	Storage modulus
E''	Loss modulus
$\mathscr{E}$	Electric field
ε	Mechanical strain, Emissivity
f	Frequency
F	Force
G	Thermal conductance
h	Height
I	Electric current
I_z	Geometrical moment of inertia in z-direction
k	Spring constant
k_B	Boltzmann constant
L	Length
$\mathscr{L}$	Electric induction
λ_n	Relative wavelength of mode n
m	Lump mass

m_0	Total mass
m_{eff}	Effective mass
M_m	Molar mass
μ	Viscosity
n	Mode number
p	Pressure
ν	Poisson's ratio
π	Mathematical constant π
Q	Quality factor
$\mathcal{Q}$	Electric charge
R_{gas}	Universal molar gas constant
$\mathcal{R}$	Electric resistance
ρ	Mass density
σ	Stress
σ_z^2	Variance of vibrational amplitude
σ_φ^2	Variance of phase
σ_y^2	Allan variance
u	Time-dependent displacement
U	Time- and position-dependent displacement
u_0	Absolute vibrational amplitude
V	Voltage
S	Power spectral density
τ	Decay time
t	Time
T	Temperature
τ	Time constant
τ_r	Resonator time constant
τ_{th}	Thermal time constant
v	Velocity and Volume
ϕ_n	Normalized time-independent displacement function
φ	Phase
w	Width
W	Energy
x	Cartesian coordinate
y	Cartesian coordinate or fractional frequency change
z	Cartesian coordinate
z_0	Vibrational amplitude of resonator
z_r	Vibrational amplitude at resonance
ζ	Damping ratio
ω	Angular frequency
ω_0	Eigenfrequency
ω_r	Resonance frequency

Chapter 1
Lumped-Element Model Resonators

At the *eigenfrequency* of an ideal (lossless) mechanical structure, according to the equipartition theorem, the kinetic energy of a specific mechanical vibration is equal to the potential energy stored in the respective vibrational deformation of the structure. Continuum mechanical structures have many such vibrational modes, called *eigenmodes*, at which this situation occurs. At the eigenfrequency, the total energy in the mechanical system is passed back and forth endlessly between kinetic and potential energy. Thus, once energy is added, e.g., by a kick, such a system would endlessly oscillate precisely at its eigenfrequency with a constant vibrational amplitude.

In a real mechanical structure however, not the entire energy is commuting between kinetic and potential energy, but a little part of the energy is lost during every cycle of vibration. There are many different mechanisms by which energy can dissipate, as will be discussed in Chap. 3. Hence, a real mechanical structure will oscillate only for a finite amount of time, until all the energy that was initially entered into the system, e.g., by the kick, has been lost. In real mechanical structures with inherent energy loss, the eigenmode mechanism is called *resonance*. And the predominant frequency at which the system moves periodically is called the *resonance frequency*. The resonance frequency is typically close (slightly lower) to the eigenfrequency of the same system assumed without losses.

The resonance frequencies of a micro- or nanomechanical resonator are typically estimated from the respective eigenfrequencies, which can be calculated by means of analytical continuum mechanical models, as will be discussed in Chap. 2, or finite element method (FEM) simulations. FEM tools are readily available and are the most convenient way to obtain a precise estimate of the eigenmodes of specific micro- and nanomechanical resonators. On the one hand, compared to FEM, analytical models represent idealized mechanical structures and often do not reproduce the specific features of the mechanical resonator to be modeled. On the other hand, analytical models give valuable insight into the effect of geometry and material parameters on the resulting eigenmodes (and eigenfrequencies). Therefore,

S. Schmid et al., *Fundamentals of Nanomechanical Resonators*,
https://doi.org/10.1007/978-3-031-29628-4_1

despite FEM, analytical continuum mechanical models are indispensable for the design process of micro- and nanomechanical resonators.

As analytical continuum mechanical models, for simplicity, typically do not consider energy losses, it is convenient to represent each individual eigenmode of a mechanical resonator by an equivalent lumped-element model, as discussed in this chapter. By using lumped-element models, it is straightforward to include energy losses, which allows for a more accurate description of the dynamic response of an individual eigenmode of a real dissipative micro- and nanomechanical resonator.

In this section, the behavior of 1-DOF (degree of freedom) mechanical resonators is discussed by means of lumped-element models. In Sect. 1.1 behavior of nonconservative linear resonators is introduced, that is, resonators that dissipate energy during vibration. Section 1.2 presents an analysis of linear resonators that are coupled to each other, i.e., 2-DOF system. Section 1.3 discusses the nonlinear behavior of resonators when driven at large amplitudes. Finally, in Sect. 1.4 the phenomenon of parametric amplification is introduced.

1.1   Damped Linear Resonator

Mechanical vibrations in physical systems are associated with the periodic conversion of kinetic to potential energy and vice versa. Elastic elements such as beams, strings, and plates are able to store potential energy in terms of the deformation energy. This is typically modeled as a spring, as shown in Fig. 1.1. If the system comes out of its position of rest, restoring forces accelerate the mass of the system toward the rest position, thereby generating kinetic energy. Due to the kinetic energy, the physical system traverses the position of rest, generating new deflection energy that is potential energy. This periodic energy conversion would continue forever if the present mechanical energy would not be resisted by dissipative forces.

For the calculation of the free vibration (eigenfrequency) of continuum mechanical resonators, as discussed in Chap. 2, dissipative forces are dismissed, and thus the total energy remains constant at any time. In a real vibrational system, energy is always dissipated by viscous damping, acoustic transmission, surface losses, internal dissipation, etc. The damping mechanisms in nanomechanical resonators are discussed in Chap. 3. Dissipative forces are approximately proportional to the velocity of the vibrational movement and are often modeled as a dashpot, as shown in Fig. 1.1.

The forced vibration of a damped system with a single degree of freedom can be modeled as a 1-DOF resonator based on a linear zero mass spring, a linear damping element, and a mass, as depicted in Fig. 1.1. Assuming a periodic driving force $F(t) = F_0 \cos(\omega t)$, the second-order differential equation of the system is

$$m\ddot{z} + c\dot{z} + kz = F(t) \tag{1.1}$$

where m is the mass, k is the spring constant and c is the damping coefficient.

Fig. 1.1 Damped
one-dimensional resonator
with one degree of freedom

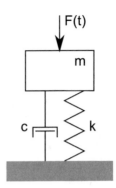

1.1.1 Free Undamped Vibration

In the specific case of a nondriven lumped-element resonator with zero damping c, the total energy of the system remains constant. In this case the resonator turns into an oscillator with a vibrational amplitude that can be expressed as $z(t) = z_0 \cos(\omega_0 t)$. During oscillation the total energy is fully swapped back and forth between kinetic (at the zero crossings) and potential (at the reversal points) energy. According to the equipartition theorem, the two energies have to be equal:

$$\frac{1}{2} m \dot{z}^2 = \frac{1}{2} k z^2 \tag{1.2}$$

which yields the *eigenfrequency* ω_0 of the undamped free mechanical system, also called an *oscillator*

$$\omega_0 = \sqrt{\frac{k}{m}}. \tag{1.3}$$

1.1.2 Free Damped Vibration

In the case of a nondriven system, the equation of motion (1.1) reduces to the homogeneous differential equation:

$$\ddot{z} + 2\zeta \omega_0 \dot{z} + \omega_0^2 z = 0 \tag{1.4}$$

with the *damping ratio*

$$\zeta = \frac{c}{2\sqrt{k\,m}}. \tag{1.5}$$

Inserting a trial solution $z(t) = z_0 e^{\gamma t}$ in (1.4) gives then solutions that satisfy

$$\gamma_{1,2} = -\omega_0 \left(\zeta \pm \sqrt{\zeta^2 - 1} \right). \tag{1.6}$$

If $\zeta > 1$, γ is real and the solution of (1.4) is an exponential decay. In this situation the system is heavily damped (overdamped case). For the special case when $\zeta = 1$, the system is critically damped. As the name implies, the damping ratio $\zeta = c/c_c$ presents the level of a system's damping c relative to critical damping c_c. Based on (1.5) the critical damping coefficient obviously is given by $c_c = 2\sqrt{k\,m}$.

The system behavior becomes more interesting when $\zeta < 1$, i.e., when $\Im(\gamma_{1,2}) \neq 0$. In that case, the system is slightly damped or underdamped and (1.6) can be written as

$$\gamma_{1,2} = -\omega_0 \left(\zeta \pm i\sqrt{1 - \zeta^2} \right). \tag{1.7}$$

For a linear homogeneous ordinary differential equation, the linear sum of linearly independent solutions $\gamma_{1,2}$ are also solutions. Then, the solution of the homogeneous differential equation (1.4) is

$$z(t) = e^{-\omega_0 \zeta t} \left[z_1 e^{-i\omega_0 \sqrt{1-\zeta^2}t} + z_2 e^{i\omega_0 \sqrt{1-\zeta^2}t} \right]. \tag{1.8}$$

Let's solve (1.8) for the case that the mass is initially displaced by z_0 and then let go. These initial conditions $z(0) = z_0$ and $\dot{z}(t) = 0$ yield values for the partial amplitudes $z_1 = z_2 \approx z_0/2$ for $\zeta \ll 1$. With these values, and by applying Euler's formula, the solution of (1.4) for these specific conditions becomes:

$$z(t) = z_0 e^{-\omega_0 \zeta t} \cos\left(\omega_0 \sqrt{1 - \zeta^2}t \right). \tag{1.9}$$

This solution is plotted in Fig. 1.2 for an initial displacement $z_0 = 1$ and $\zeta = 0.03$. The plot indicates that the system has an exponentially decaying oscillation with a frequency $\omega_{nat} = \omega_0\sqrt{1 - \zeta^2}$, which in this context is also called *natural frequency*.

Fig. 1.2 Plot of (1.9) for $z_0 = 1$, $\omega_0 = 1$, and $\zeta = 0.03$

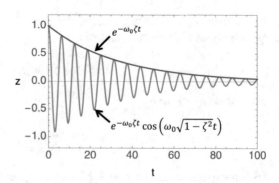

1.1.3   Driven Damped Vibration

Now we study the response of a linear lumped-element resonator to a harmonic driving force $F(t) = F_0 \cos(\omega t + \Delta)$ with an arbitrary phase Δ. In this case, the system will move with a combination of the transient response from (1.9) and a steady-state response. For the derivation of the steady-state response, the absolute phase Δ is irrelevant and can be omitted. The differential equation of motion (1.1) can be solved elegantly by using complex variables for the driving force $\hat{F}(t) = F_0 e^{i\omega t}$ and amplitude $\hat{z}(t) = \hat{z}_0 e^{i\omega t}$. Inserting these complex variables into (1.1) yields the complex amplitude:

$$\hat{z}_0 = \frac{F_0/m}{(\omega_0^2 - \omega^2) + 2i\zeta \omega_0 \omega} = z_0 e^{i\varphi}, \tag{1.10}$$

which can be converted to the polar form with the norm

$$z_0 = |\hat{z}_0| = \frac{F_0/m}{\sqrt{(\omega_0^2 - \omega^2)^2 + 4\zeta^2 \omega_0^2 \omega^2}} \tag{1.11}$$

and the phase

$$\varphi = arg(\hat{z}_0) = \arctan \frac{2\zeta \omega_0 \omega}{\omega^2 - \omega_0^2}. \tag{1.12}$$

The solution then takes the form

$$z(t) = \Re\left(\hat{z}(t)\right) = \Re\left(z_0 e^{i\varphi} e^{i\omega t}\right)$$
$$= \frac{F_0/m}{\sqrt{(\omega_0^2 - \omega^2)^2 + 4\zeta^2 \omega_0^2 \omega^2}} \cos(\omega t + \varphi). \tag{1.13}$$

Both the amplitude and phase response can be written as a function of the relative frequency ω/ω_0:

$$z_0 = \frac{F_0/k}{\sqrt{\left(1 - \left(\frac{\omega}{\omega_0}\right)^2\right)^2 + 4\zeta^2 \left(\frac{\omega}{\omega_0}\right)^2}}. \tag{1.14}$$

and

$$\varphi = \arctan \frac{2\zeta \left(\frac{\omega}{\omega_0}\right)}{\left(\frac{\omega}{\omega_0}\right)^2 - 1}. \tag{1.15}$$

The numerator in (1.14) $z_s = F_0/k$ represents the static deflection. Hence, the displacement can be written as

$$z(t) = z_s \, \delta z_0 \cos(\omega t + \varphi) \tag{1.16}$$

with the relative amplitude response δz_0 (also called *dynamic amplification* or *gain*) given by

$$\delta z_0 = \frac{1}{\sqrt{\left(1 - \left(\frac{\omega}{\omega_0}\right)^2\right)^2 + 4\zeta^2 \left(\frac{\omega}{\omega_0}\right)^2}}. \tag{1.17}$$

Both the amplitude and phase response of a linear resonator are plotted in Fig. 1.3b and c.

Another common way to actuate a mechanical resonator is by shaking its base by, e.g., a piezoelectric shaker. The schematic representation of this scenario is shown in Fig. 1.3d. In this case, since we do not apply a force directly but rather an acceleration, the relative amplitude and phase response become [1]:

$$\delta z_0 = \frac{\left(\frac{\omega}{\omega_0}\right)^2}{\sqrt{(1 - \left(\frac{\omega}{\omega_0}\right)^2)^2 + 4\zeta^2 \left(\frac{\omega}{\omega_0}\right)^2}}. \tag{1.18}$$

and

$$\varphi = \arctan \frac{2\zeta \left(\frac{\omega}{\omega_0}\right)}{1 - \left(\frac{\omega}{\omega_0}\right)^2}. \tag{1.19}$$

Both functions (1.18) and (1.19) are plotted in Fig. 1.3e and f.

In the case driven by a force, from (1.17) it can be seen that the amplitude of a slightly damped system has a maximum near the eigenfrequency ω_0, the so-called resonance peak. The frequency of the highest amplitude is called *resonance frequency* ω_r, and it can be found at $\partial z_0/\partial \omega = 0$. The resonance frequency is given by

$$\omega_r = \omega_0 \sqrt{1 - 2\zeta^2}. \tag{1.20}$$

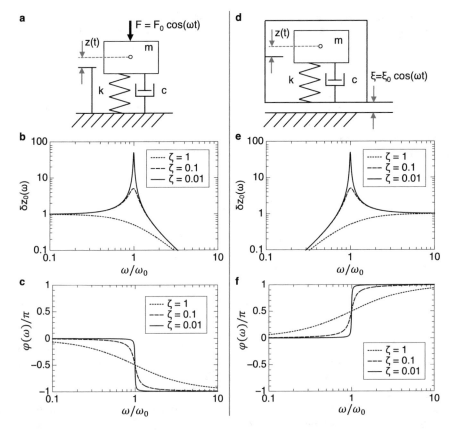

Fig. 1.3 Driven damped vibration of a lumped system. Schematic drawings of a damped resonator (**a**) driven by a force and (**d**) driven by an external vibration. (**b**, and **e**) show the respective relative amplitude responses (1.17) and (1.18), and (**c** and **f**) show the respective phase responses (1.15) and (1.19)

For slight damping ($\zeta \ll 1$), the resonance frequency is very close to the eigenfrequency of the system $\omega_r \approx \omega_0$.

When inserting ω_r into (1.15), the phase lag at resonance becomes:

$$\varphi_r = \arctan \frac{\sqrt{1 - 2\zeta^2}}{\zeta}. \tag{1.21}$$

For slight damping, the phase lag at resonance approaches $\varphi_r \approx \arctan(1/\zeta) = \pi/2$.

At resonance, the vibrational amplitude (1.14) reaches a maximal value of

$$z_r = z_0\big|_{\omega=\omega_r} = \frac{F_0}{k} \frac{1}{2\zeta\sqrt{1 - \zeta^2}} = z_s Q = \frac{F_0}{k} Q. \tag{1.22}$$

The maximal amplitude at resonance is the static deflection times an amplification factor Q, which is called the *quality factor*, which for slight damping becomes

$$Q \approx \frac{1}{2\zeta} \quad (\text{as } \zeta \ll 1). \tag{1.23}$$

From the frequency response shown in Fig. 1.3, it can further be observed that the slope of the phase response at resonance also becomes steeper with lower damping. Indeed, the quality factor can also be determined from the slope of the phase response (1.15) at resonance, assuming small damping $\omega_r \approx \omega_0$

$$\left.\frac{\partial \varphi}{\partial \omega}\right|_{\omega=\omega_0} = \frac{1}{\zeta\omega_0} \approx \frac{2Q}{\omega_0}. \tag{1.24}$$

In summary, the differential equation of motion of a driven linear resonator (1.1) with slight damping ($\zeta \ll 1$) can be written as a function of the two key characteristics ω_0 and Q of a resonator:

$$\ddot{z}(t) + \frac{\omega_0}{Q}\dot{z}(t) + \omega_0^2 z(t) = \frac{F(t)}{m}. \tag{1.25}$$

ω_0 and Q of typical nanomechanical resonators are discussed in Chaps. 2 and 3, respectively. Chapter 4 presents methods on how to produce the driving force $F(t)$ and how to observe $z(t)$. In Chap. 6, the response of displacement z_0, eigenfrequency ω_0, and quality factor Q as a function of $F(t)$ and the resonator parameters m, k, and c are discussed. And finally, Chap. 5 discusses the situation when the coherent driving force is replaced by thermal white noise $S_F(\omega)$ and its implications on the precision with which amplitude and resonance frequency can be detected.

1.1.3.1 Quality Factor

Besides the previous definition of the quality factor as amplification factor of the vibration of a linear resonator at resonance (1.22), there are other definitions that lead to the expression for Q under the assumption of slight damping. Here, two other definitions are introduced. The first definition is a physical consideration based on energy dissipation, and the second definition is based on the peak sharpness.

Ring Down

The physical definition is the ratio between the energy stored and energy lost during one cycle at resonance:

$$Q = 2\pi \frac{W}{\Delta W} \tag{1.26}$$

where W is the total energy stored in the system and ΔW is the energy loss during one cycle of oscillation. With a displacement of $z = z_0 \cos \omega_r t$ at resonance, the total energy of the system is

$$W = \max \left\{ \frac{1}{2} m \dot{z}^2 \right\} = \frac{1}{2} m z_0^2 \omega_r^2. \tag{1.27}$$

With the dissipative force $F_d = -c\dot{z}$, the energy lost during one cycle can be calculated as

$$\Delta W = - \int_0^{2\pi/\omega_r} F_d \dot{z} \, dt = \int_0^{2\pi/\omega_r} c\dot{z}^2 \, dt = \pi c z_0^2 \omega_r \tag{1.28}$$

and

$$Q = 2\pi \frac{W}{\Delta W} = \frac{m\omega_r}{c}. \tag{1.29}$$

With (1.5) and (1.20), Q can be written as

$$Q = \frac{\sqrt{1 - 2\zeta^2}}{2\zeta}, \tag{1.30}$$

which for slight damping reduces to (1.23).

The energy dissipation consideration can be directly observed from the ring down of the vibration after the driving force has stopped. In this situation, the system can be modeled as a free damped vibration as discussed in Sect. 1.1.2, with the transient solution (1.9). The quality factor can now be found by fitting an exponential function $\exp(-t/\tau_R)$ to the envelope of the decaying vibrational amplitude, as shown in Fig. 1.2. With $\tau_R^{-1} = \zeta \omega_0$, Q (1.23) is given by

$$Q \approx \frac{1}{2} \omega_0 \tau_R. \tag{1.31}$$

The observation of Q from the ring down is particularly useful for very high Qs where, e.g., the phase angle (1.24) becomes too steep to be determined precisely and the peak width, which will be discussed next, becomes too narrow.

Full Width at Half Maximum

This method is based on the definition of Q in electrical resonant circuits where the quality factor is given by Bao [1]

Fig. 1.4 Comparison of the power response of a linear resonator to the Lorentzian function

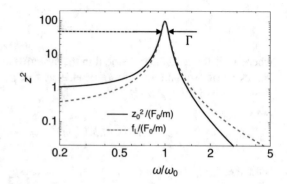

$$Q = \frac{\omega_r}{\Gamma} = \frac{1 - 2\zeta^2}{2\zeta} \tag{1.32}$$

where Γ is the frequency difference between the two frequencies at which the amplitude curve (1.14) has the half maximum energy, which is at $z_0(\omega_r)/\sqrt{2}$. It's also called the *full width at half maximum* (FWHM). For slight damping ($\zeta \ll 1$), this definition of the quality factor is equal to the physical definition (1.30), and the FWHM can be approximated by

$$\Gamma \approx 2\zeta\omega_0 \quad (\text{as } \zeta \ll 1). \tag{1.33}$$

The quality factor can now be found by measuring the amplitude response around the resonance. According to (1.32) the resonance frequency divided by the -3dB bandwidth (Γ) is an approximation for Q for slight damping.

In a typical electrical measurement, the transfer function is measured in units of power instead of amplitude. The power is proportional to the square of the resonator amplitude (1.14). For slightly damped systems, the resulting response (1.11) can well be approximated by the Lorentzian function f_L, which is related to the Cauchy distribution:

$$z_0^2 \approx f_L(\omega) = \frac{(F_0/m)^2}{4\omega_0^2} \left[\frac{1}{(\omega_0 - \omega)^2 + \left(\frac{1}{2}\Gamma\right)^2} \right]. \tag{1.34}$$

A comparison of the power response of a linear resonator to the corresponding *Lorentzian* (1.34) is shown in Fig. 1.4.

1.1.3.2 Transient Response

The transient response of a damped resonator in the case of a suddenly ceasing driving force, causing the free vibration of a resonator, has been derived in

Fig. 1.5 Plot of (1.37) for $z_s = 1$, $\omega_0 = 1$, and $Q = 30$

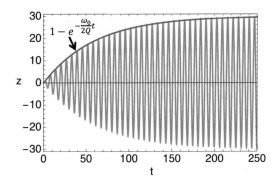

Sect. 1.1.2. It was shown that the free oscillation amplitude is decaying with a time constant $\tau_R = 1/(\omega_0 \zeta)$. In this section the transient response of a driven resonator is discussed by means of solving the general solution for specific initial conditions. The general solution of the differential equation of motion (1.25) is made up of the homogeneous (1.8) and the particular solution (1.16):

$$z(t) = e^{-\omega_0 \zeta t}\left[z_1 e^{-i\omega_0 \sqrt{1-\zeta^2}t} + z_2 e^{i\omega_0 \sqrt{1-\zeta^2}t} \right] + (F_0/k)\,\delta z_0 \cos(\omega t + \varphi). \tag{1.35}$$

Let's discuss the case of a slightly damped ($\zeta \ll 1$) resonator at rest that is driven at its resonance frequency ($\omega = \omega_0 \approx \omega_r$) with an oscillating force, and then (1.35) simplifies to

$$z(t) = e^{-\frac{\omega_0}{2Q}t}\left[z_1 e^{-i\omega_0 t} + z_2 e^{i\omega_0 t} \right] + z_s\, Q \cos(\omega_0 t + \varphi) \tag{1.36}$$

with the quasi-static displacement $z_s = F_0/k$. The initial conditions $z(0) = \dot{z}(0) = 0$ produce a system of equations that can be solved for z_1 and z_2. With the known phase at resonance $\varphi = -\pi/2$, the two partial amplitudes become $z_1 = -iQz_s/2$ and $z_2 = iQz_s/2$, and the displacement function (1.36) reduces to

$$z(t) = \left(1 - e^{-\frac{\omega_0}{2Q}t}\right) z_s\, Q \sin(\omega_0 t). \tag{1.37}$$

This response is plotted in Fig. 1.5, showing how the displacement builds up until the maximum amplitude is reached. This response represents the quintessential depiction of the resonance phenomena, where a small oscillating force can build up a large vibrational amplitude if the driving frequency matches the resonance frequency of the system. In summary, a resonator responds to a change of the driving force (buildup or ring-down) with a time constant $\tau_R = 2Q/\omega_0$.

Next, let's discuss the case of an abrupt change of the driving force frequency relative to the resonance frequency. This situation can happen in two different scenarios. First, the resonator parameters remain unchanged and the driving force frequency changes. Second, the driving force frequency remains constant, and the eigenfrequency of the resonator changes due to a change of a resonator parameter. Both these cases can be understood as the superposition of a ring-down and a build-up response. At the moment the relative frequency change happens, on the one hand, the resonator vibration rings down (1.9) because the driving force at the initial frequency ceased, and on the other hand, the resonator vibration builds up (1.37) because of the driving force at the new frequency. Both the ring-down and the buildup have the same time constant $\tau_R = 1/(\omega_0 \zeta)$, with which such a frequency transition happens [2].

1.2 Coupled Linear Resonators

Coupled mechanical resonators are, e.g., studied for the modeling of mechanical waves in crystalline materials, where the individual atoms are represented by lumped masses that are coupled with each other via linear springs. Besides such more theoretical applications, the rich dynamics of mutually coupled micro- and nanomechanical resonators have been explored for various applications. Mechanically coupled resonators have, e.g., been used for mass sensing with micromechanical resonator pairs[3, 4] or arrays [5]. The coupling can also happen via electrostatic forces [6], which allow the design of bandpass filters with variable filter properties [7]. Recently, in optomechanics, mechanical resonators have been coupled to RF or optical cavities [8]. In this section, a simple case of mutually coupled mechanical resonators with equal mass and spring constant is presented. A real-world application of such a coupled micromechanical resonator pair is shown in Fig. 1.6. In this example, the two cantilevers which vibrate out of plane are coupled via a shared overhang between the structures.

A more detailed introduction to strong coupling and the involved so-called avoided crossing is presented by Novotny [9]. Besides the coupling of two mechanical resonators, it's possible to couple resonant systems from different physical domains. A system with two mechanical resonators that are mechanically coupled is schematically depicted in Fig. 1.7.

symmetric mode   antisymmetric mode

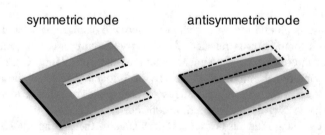

Fig. 1.6 Schematic visualization of the symmetric and antisymmetric mode of vibration of two mechanically coupled cantilevers

Fig. 1.7 Schematic drawing
of two coupled linear damped
resonators

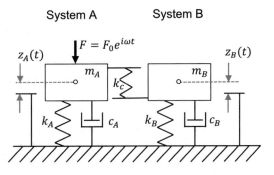

1.2.1   Coupled Undamped Resonators

Equal oscillators that are coupled are the most simple coupled system that can be studied. In this case, the coupled system in Fig. 1.7 is described by ($m_A = m_B = m$ & $k_A = k_B = k$). Further, it is undamped ($c_A = c_B = 0$) and undriven ($F(t) = 0$). Newton's second law readily yields the equations of motion:

$$m\ddot{z}_A + kz_A + k_c(z_A - z_B) = 0$$
$$m\ddot{z}_B + kz_B - k_c(z_A - z_B) = 0. \tag{1.38}$$

This linear system of equations can be solved by plugging in the standard ansatz:

$$\begin{pmatrix} z_A(t) \\ z_B(t) \end{pmatrix} = \begin{pmatrix} C_A \\ C_B \end{pmatrix} \cos(\omega t), \tag{1.39}$$

which yields the linear system of equations

$$\begin{pmatrix} -\omega^2 m + k + k_c & -k_c \\ -k_c & -\omega^2 m + k + k_c \end{pmatrix} \begin{pmatrix} C_A \\ C_B \end{pmatrix} = \begin{pmatrix} 0 \\ 0 \end{pmatrix}. \tag{1.40}$$

The nontrivial solutions can be found when the determinant of the system is zero:

$$(-\omega^2 m + k + k_c)^2 - k_c^2 = 0, \tag{1.41}$$

which yields two normal eigenfrequencies

$$\omega_- = \sqrt{\frac{k}{m}} \qquad \omega_+ = \sqrt{\frac{k + 2k_c}{m}}, \tag{1.42}$$

which are the symmetric and antisymmetric eigenfrequency, respectively.

Plugging in these eigenfrequencies in (1.40) gives $C_A = C_B$ for ω_- and $C_A = -C_B$ for ω_+. For the symmetric mode, with eigenfrequency ω_-, the vibrational amplitudes of both resonators are equal. Its eigenfrequency is equal to the eigenfrequency of a single resonator in the absence of coupling ($\omega_- = \omega_0$). Since both resonators are vibrating in phase, the coupling spring k_c remains unstretched. For the antisymmetric mode, with eigenfrequency ω_+, the two resonators move in opposite direction, and the coupling spring is under a lot of strain, which increases the normal mode eigenfrequency.

Now let's study the response of this simple system once it's been set in motion. The solution of the coupled system is a superposition of its normal modes. The following derivation is performed with the real part of the complex ansatz (1.39). The displacement response of the system is then described by

$$
\begin{aligned}
z_A &= z_- \cos\left(\omega_- t\right) + z_+ \cos\left(\omega_+ t\right) \\
z_B &= z_- \cos\left(\omega_- t\right) - z_+ \cos\left(\omega_+ t\right).
\end{aligned}
\tag{1.43}
$$

Here, z_- and z_+ are the partial displacements of the symmetric and antisymmetric mode, respectively. Since the vibrational amplitudes of the symmetric mode are in phase with each other ($C_A/C_B = +1$), both partial amplitudes z_- in 1.43 are positive. In contrast, the partial antisymmetric amplitudes of systems A and B are exactly 180° out of phase to each other ($C_A/C_B = -1$), and therefore z_+ of system B has a negative sign.

In the case discussed here, it is assumed that at $t = 0$ oscillator A has an arbitrary amplitude C, while oscillator B is in a position of rest:

$$
\begin{aligned}
z_A &= C = z_- + z_+ \\
z_B &= 0 = z_- - z_+.
\end{aligned}
\tag{1.44}
$$

This readily yields $z_- = z_+ = C/2$. Using the trigonometric transformations $\cos a + \cos b = 2 \cos\frac{a+b}{2} \cos\frac{a-b}{2}$ and $\cos a - \cos b = -2 \sin\frac{a+b}{2} \sin\frac{a-b}{2}$, (1.43) can be written as

$$
\begin{aligned}
z_A(t) &= C \cos\left(\frac{\omega_+ + \omega_-}{2} t\right) \cos\left(\frac{\omega_+ - \omega_-}{2} t\right) \\
z_B(t) &= C \sin\left(\frac{\omega_+ + \omega_-}{2} t\right) \sin\left(\frac{\omega_+ - \omega_-}{2} t\right).
\end{aligned}
\tag{1.45}
$$

In the case that the coupling spring constant is smaller than the spring constant ($k_c < k$), the two modes (1.43) have similar frequencies whose interference produces a beat pattern, as plotted in Fig. 1.8. This is obvious from (1.45) where the first term of each system represents the "fast" oscillation frequency ($\omega_+ + \omega_-)/2$. And the second term produces a "slow" signal envelope with ($\omega_+ - \omega_-)/2$. The plot in Fig. 1.8 shows how the entire system energy is completely swapped between system A and system B every quarter beat period with the characteristic time:

Fig. 1.8 Response of two coupled oscillators (1.45) for the starting conditions $z_A = C$ and $z_B = 0$

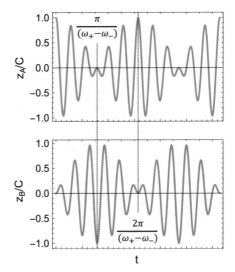

$$\tau_c = \frac{\pi}{\omega_+ - \omega_-}. \tag{1.46}$$

According to (1.42), the larger the coupling strength k_c, the larger the frequency difference between the two normal modes, and the faster is the energy exchange between the two oscillators.

1.2.1.1 Avoided Crossing

Here, the behavior of two different resonators is discussed. For simplicity, both systems A and B in Fig. 1.7 are considered to be undamped ($c_A = c_B = 0$) and undriven ($F = 0$). Similar to the previous section, Newton's second law readily yields the equations of motion:

$$m_A \ddot{z}_A + k_A z_A + k_c(z_A - z_B) = 0$$
$$m_B \ddot{z}_B + k_B z_B - k_c(z_A - z_B) = 0. \tag{1.47}$$

This linear system of equations can be solved by plugging in the standard ansatz (1.39). As before, the nontrivial solutions for this homogeneous system of equations can be found when the determinant becomes zero:

$$\det \begin{pmatrix} -\omega^2 m_A + k_A + k_c & -k_c \\ -k_c & -\omega^2 m_B + k_B + k_c \end{pmatrix} = 0, \tag{1.48}$$

which yields the two positive normal eigenfrequencies [9]

$$\omega_{\pm}^2 = \frac{1}{2}\left(\omega_A^2 + \omega_B^2 \pm \sqrt{(\omega_A^2 - \omega_B^2)^2 + 4\Gamma_c^2\omega_A\omega_B}\right) \tag{1.49}$$

with $\omega_A = \sqrt{(k_A + k_c)/m_A}$ and $\omega_B = \sqrt{(k_b + k_c)/m_B}$, and

$$\Gamma_c = \frac{\sqrt{k_c/m_A}\sqrt{k_c/m_B}}{\sqrt{\omega_A\omega_B}}. \tag{1.50}$$

It can readily be verified that (1.49) reduces to (1.42) for $k_A = k_B = k$ and $m_A = m_B = m$.

In order to visualize the eigenfrequencies $\omega_{\pm}$, system A is kept with a constant k_A and m_A, while the uncoupled eigenfrequency of system B is tuned by changing the stiffness k_B but keeping the mass m_B constant. The normal eigenfrequencies (1.49) are plotted in Fig. 1.9. For the case without coupling ($k_c = 0$), the independent eigenfrequencies of the two systems intersect at $\Delta k = k_A - k_B = 0$. In contrast, for the case of coupling ($\Delta k = 0.1k$), a characteristic *avoided crossing* or *anticrossing* can be observed. In this case where $m_A = m_B = m$ and $k_A = k_B = k$, the magnitude of the frequency splitting is given by

$$\omega_+ - \omega_- = \Gamma_c \approx \frac{k_c}{m\omega_0} = \frac{k_c}{k}\omega_0 \quad (\text{if } k_c \ll k). \tag{1.51}$$

The frequency splitting Γ_c scales with the coupling spring constant k_c and hence is a measure of the coupling strength.

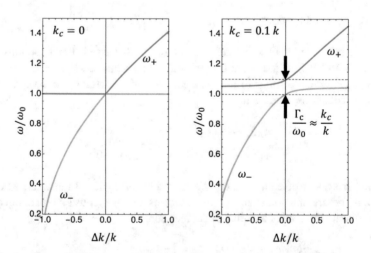

Fig. 1.9 Plot of the eigenfrequencies (1.49) in relation to $\omega_0 = \sqrt{k/m}$. The parameters of system A are kept constant ($k_A = k$ & $m_A = m$), while the parameters of system B are tuned ($k_B = k + \Delta k$ & $m_B = m$)

1.2.2 Driven Coupled Resonators

Finally, damping is introduced into the analysis. The response of a damped and driven coupled system, as depicted in Fig. 1.7, for two equal resonators ($m_A = m_B = m$ and $k_A = k_B = k$) can be calculated from the corresponding system of equations by using complex variables:

$$m\ddot{z}_A + c_A\dot{z}_A + kz_A + k_c(z_A - z_B) = F_0\,e^{i\omega t}$$
$$m\ddot{z}_B + c_B\dot{z}_B + kz_B - k_c(z_A - z_B) = 0. \tag{1.52}$$

Assuming a heavily underdamped system ($\zeta \ll 1$), with (1.3), (1.33), and (1.50), these equations can be rewritten in the following way:

$$\ddot{z}_A + \Gamma_A\dot{z}_A + (\omega_0^2 + \omega_0\Gamma_c)z_A - \omega_0\Gamma_c z_B = \frac{F_0}{m}e^{i\omega t}$$
$$\ddot{z}_B + \Gamma_B\dot{z}_B + (\omega_0^2 + \omega_0\Gamma_c)z_B - \omega_0\Gamma_c z_A = 0. \tag{1.53}$$

This system of equations can readily be solved by inserting the complex ansatz $z_A(t) = C_A e^{i\omega t}$ and $z_B(t) = C_B e^{i\omega t}$ to (1.53), which results in the complex amplitude response functions:

$$z_A(t) = \frac{(\omega_0^2 + \omega_0\Gamma_c - \omega^2 + i\omega\Gamma_B)(F_0/m)e^{i\omega t}}{(\omega_0^2 + \omega_0\Gamma_c - \omega^2 + i\omega\Gamma_A)(\omega_0^2 + \omega_0\Gamma_c - \omega^2 + i\omega\Gamma_B) - (\omega_0\Gamma_c)^2}$$
$$z_B(t) = \frac{(\omega_0\Gamma_c)(F_0/m)e^{i\omega t}}{(\omega_0^2 + \omega_0\Gamma_c - \omega^2 + i\omega\Gamma_A)(\omega_0^2 + \omega_0\Gamma_c - \omega^2 + i\omega\Gamma_B) - (\omega_0\Gamma_c)^2}. \tag{1.54}$$

Figure 1.10 shows the coupled system response (1.54) for the case when both resonators have equal damping $\Gamma_A = \Gamma_B = \Gamma$. It can be seen that the normal mode splitting becomes observable for $\Gamma_c \geq \Gamma$. This regime is also called the *strong coupling* regime. On the one hand, according to (1.46), the frequency splitting Γ_c is a direct measure of the energy transfer rate between the resonators, $\tau_c^{-1} = \Gamma_c/\pi$. On the other hand, as discussed in Sect. 1.1.3.1, Γ is a direct measure of the energy dissipation rate $\tau_R^{-1} = \Gamma/2$. That means that in the strong coupling regime, the energy transfer between systems A and B happens faster than the energy can dissipate in either of the two systems.

Figure 1.11 shows the coupled system response (1.54) for the case when the damping of the passive resonator B is lower than the driven resonator A ($\Gamma_B < \Gamma_A$). Here two interesting phenomena can be observed. First, with increasing coupling strength, the FWHM Γ_B of the lower-loss system B starts to match Γ_A of the lossy system A. And second, for small coupling, resonator B is drawing energy from resonator A, which results in a sharp and thin dip in the response of resonator A at ω_0. This dip is related to a phenomenon in optics called *electromagnetically induced*

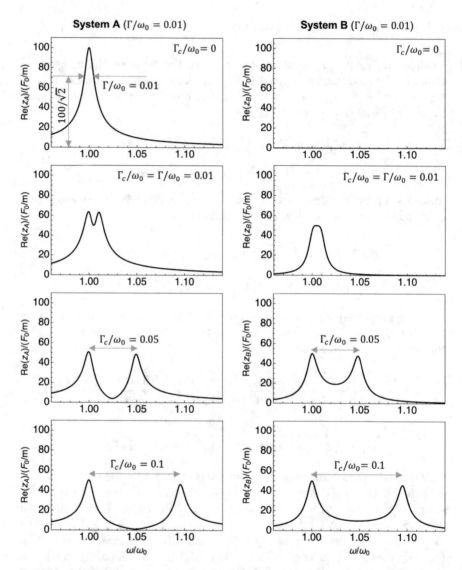

Fig. 1.10 Plot of the frequency response (1.54) of both, the driven system A and the passive system B, for an increasing coupling strength Γ_c. Both resonators A and B have a relative FWHM of $\Gamma_A = \Gamma_B = 0.01\omega_0$

transparency (EIT) [10]. The situation that the passive system B has lower damping than the driven system is often met when electromagnetic resonators are coupled to high-Q micro- and nanomechanical resonators where such *optomechanically induced transparencies* have been observed [11–13].

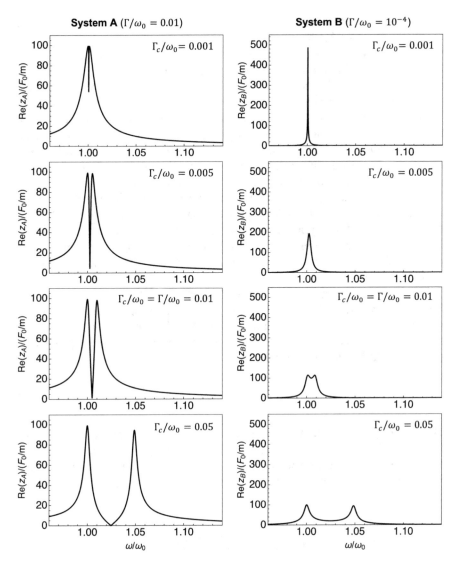

Fig. 1.11 Plot of the frequency response (1.54) of the driven system A with $\Gamma_A = 0.01\omega_0$) and the passive system B with $\Gamma_B = 10^{-4}\omega_0$), for an increasing coupling strength Γ_c

1.3 Damped Nonlinear Resonators

In previous sections, the behavior of mechanical resonators and their lumped-element models has been analyzed within what is typically called the linear regime. This stands for the regime where the amplitude of the resonator enters linearly in the equation of motion or, in other words, that the relation between the applied force and the response of the resonator (at a given frequency) does not change when the amplitude of motion becomes larger.

Unfortunately, reality is far away from this simplistic point of view, and nonlinearities are often encountered in most physical systems [14, 15]. The aim of this section is to solve the resulting equation of motion and briefly introduce the different sources of nonlinearity.

Solving a differential equation like (1.1) is relatively easy, as one can use the Fourier transform to convert the differential equation into an algebraic one. On the contrary, when nonlinear terms are part of the equation, the solution becomes less trivial. In this section, the approach taken by Cross and Lifshitz [14] to solve the nonlinear equation will be followed, and it is based on assuming that the nonlinearity is a small perturbation to the linear case. This implies that the response of the system to a harmonic external drive will be given by a fast harmonic response modulated, due to the nonlinear terms, at a much slower pace. This approximation is valid only if the quality factor of the resonator Q is large enough. Indeed, one can consider the inverse of the quality factor as the small number for which the solution is expanded around the linear case.

Here, the most common nonlinearity is discussed, in which the nonlinearity happens symmetrically in positive $+z$ and negative $-z$ vibrational amplitudes. This is fulfilled for a restoring force that scales with the cubic amplitude αz^3. In order to simplify the analysis, let us take the equation of motion of a slightly damped system (1.25) and add the additional cubic term:

$$\ddot{z}(t) + \frac{\omega_0}{Q}\dot{z}(t) + \omega_0^2 z(t) + \frac{\alpha}{m}z^3(t) = \frac{F(t)}{m}. \tag{1.55}$$

This equation is the *Duffing equation*. In addition, it is typically a good idea to work with dimensionless magnitudes. This way, the results hold validity in a universal reference frame that can be translated to actual units for each resonator independently. A normalization that is typically used consists of two steps; the first one reads as follows:

$$\hat{t} = \omega_0 t; \quad \hat{z} = z\sqrt{\frac{\alpha}{m\omega_0^2}}; \quad \hat{F} = F\sqrt{\frac{\alpha}{m^3\omega_0^6}}; \quad \hat{\omega} = \frac{\omega}{\omega_0}; \tag{1.56}$$

and this converts (1.55) into

$$\ddot{\hat{z}} + \frac{\dot{\hat{z}}}{Q} + \hat{z} + \hat{z}^3 = \hat{F}\cos\left(\hat{\omega}\hat{t}\right), \tag{1.57}$$

noting that now $\dot{\hat{z}} = \frac{d\hat{z}}{d\hat{t}}$.

The second step in the normalization is based on the fact that the system will mostly behave linearly or, in other words, that the nonlinearity is going to be weak. This latter condition is imposed by forcing the cubic term in (1.57) to be a factor proportional to Q smaller than the linear term(s). From this, it follows automatically that the driving term has to be of the order of $Q^{-3/2}$. Finally, it is possible to see that the dynamics of the mode are going to be limited to a range of frequencies around the resonance frequency, which allows the definition of the *slow* times cale and frequency. Overall

$$T = \frac{\hat{t}}{Q} = \frac{\omega_0 t}{Q}; \quad \Omega = Q(\hat{\omega} - 1) = Q(\frac{\omega}{\omega_n} - 1); \quad g = \frac{\hat{F}}{Q^{3/2}}; \quad (1.58)$$

and the amplitude will take the form:

$$x = \frac{1}{2Q^{1/2}} \left(A(T)e^{i\hat{t}} + c.c. \right) + \frac{x_1(t)}{Q^{3/2}} + \dots$$

$$\dot{x} = \frac{d\hat{x}}{d\hat{t}} = \frac{1}{2Q^{1/2}} \left(\left[iA + \frac{1}{Q}\frac{dA}{dT} \right] e^{i\hat{t}} + c.c. \right) + \frac{\dot{x}_1(t)}{Q^{3/2}} + \dots;$$

$$\ddot{x} = \frac{D^2\hat{x}}{d\hat{t}^2} = \frac{1}{2Q^{1/2}} \left(\left[-A + \frac{2i}{Q}\frac{dA}{dT} + \frac{1}{Q^2}\frac{d^2A}{dT^2} \right] e^{i\hat{t}} + c.c. \right) + \frac{\ddot{x}_1(t)}{Q^{3/2}} + \dots;$$

$$(1.59)$$

where c.c. stands for complex conjugate, in order to simplify the formulas. Due to the slow time scales, it can be seen that Eq. (1.59) has some terms scaled by different powers of the *small parameter*. By plugging (1.59) into (1.57) and picking all the terms proportional to $Q^{-3/2}$, we finally obtain an equation for the slow-varying amplitude A:

$$\frac{dA}{dT} = -\frac{1}{2}A + i\frac{3}{8}|A|^2 A - i\frac{g}{2}e^{i\Omega T}. \quad (1.60)$$

The implications and fields of application of this very general equation reach far beyond the scope of this book. The amplitude A is a complex magnitude that takes the shape: $A(T) = |A|e^{i\varphi}e^{i\Omega T}$ which should be reminded is a dimensionless representation of: $z(t) = \frac{1}{2\sqrt{Q}}\sqrt{\frac{m\omega_0^2}{\alpha}}|A| \left(e^{i\varphi}e^{i\Omega T}e^{i\omega_0 t} + c.c. \right)$. By plugging the shape of the amplitude into the *amplitude equation*, (1.60), it is possible to reach:

$$|A|^2 = \frac{g^2}{\left(2\Omega - \frac{3}{4}|A|^2 \right)^2 + \frac{1}{4Q^2}};$$

$$\tan \varphi = \left(2\Omega - \frac{3}{4}|A|^2 \right)^{-1}; \quad (1.61)$$

Figure 1.12 shows the results of Eq. (1.61), scaled amplitude of vibration (A) as a function of the scaled driving frequency (Ω) for different magnitudes of the driving force (g). It is possible to see how for small driving forces a Lorentzian curve is recovered, but as soon as the amplitude and drive increase, the curves

Fig. 1.12 Scaled amplitude of motion of a nonlinear resonator versus the scaled frequency for different driving forces

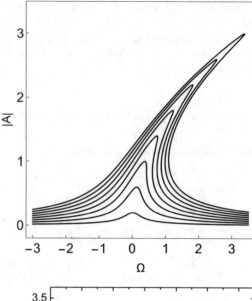

Fig. 1.13 Scaled amplitude of motion of a nonlinear resonator versus the scaled frequency for different driving forces. In this case it is shown the typical responses found experimentally, with the different result for the sweep of the frequency in the up (darker gray) or down (lighter gray) direction

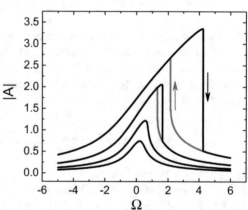

stop being symmetrical with respect to the resonant frequency, tilting to the right showing the typical behavior of a so-called Duffing resonator. Indeed, it can also be seen how after a certain drive, critical drive g_c, there are frequencies for which three amplitudes hold Eq. (1.61) true. Out of these three solutions, one can perform a stability analysis and see that one of them is unstable, whereas the other two solutions are stable. This bistability is also typical from *Duffing* resonators, and it also causes hysteresis in the resonator response when performing a frequency sweep, as depending whether you come from left to right or vice versa the final response is different, as can be seen in Fig. 1.13.

By deriving the first equation in (1.61), one can locate the frequencies for which the amplitude of response will be maximum, by imposing $\frac{d|A|^2}{d\Omega} = 0$:

$$\Omega_{max} = \frac{3}{8}|A|^2_{max}; \quad \leftrightarrow \quad \omega_{0,max} = \omega_0 + \frac{3}{8}\frac{\alpha}{m\omega_0}|z|^2_{max}; \tag{1.62}$$

which delineates one of the most used methods to determine the nonlinear coefficient, α, i.e., fitting the location in frequency of the maximum amplitude point for many actuation levels. In this case, of course, a very careful calibration of the displacement needs to be done prior to drawing any meaningful conclusion.

It is also possible to calculate the different *saddle nodes* of the resonator response, or those points at which the behavior is supposed to transition from a stable solution to an unstable one, and that in reality correspond to the vertical jumps in the amplitude response that can be seen in Fig. 1.13. Those points can be calculated by imposing $\frac{d\Omega}{d|A|^2} = 0$, which results in two families of points, one family for the transitions when the sweep is performed left to right and another family for the transitions when the sweep is performed right to left. The most interesting of all those *saddle nodes* is the one that appears the first, when both families intersect. For that point, it is also true that $\frac{d^2\Omega}{(d|A|^2)^2} = 0$, and it determines a parameter that is known as *critical amplitude*, a_c, typically considered as the threshold at which the resonator behavior starts being nonlinear:

$$a_c = \sqrt{\frac{8}{3\sqrt{3}}} \quad \leftrightarrow \quad z_c = \sqrt{\frac{8}{3\sqrt{3}}}\frac{1}{\sqrt{Q}}\sqrt{\frac{m\omega_0^2}{\alpha}}. \tag{1.63}$$

In order to avoid seeing any nonlinear effects, it is necessary to ensure that the amplitude of motion of the resonator stays well below the critical amplitude. However in practice, and for simplicity, it is considered that whenever the resonator amplitude is below z_c, linear equations can be considered. The question then is how much is this critical amplitude for a given structure. Evidently, this primarily depends on the source of nonlinearity that is limiting the device under study.

1.3.1 Sources of Nonlinearity

1.3.1.1 Material Nonlinearity

The most straightforward source of nonlinearity is the so-called material nonlinearity and happens when the relation between strain and stress is not linear anymore. Continuum solid mechanics dictates that for small deformations, stress and strain are related through the Young's modulus. However, when strain becomes larger, this ratio is modified and plasticity occurs[16]. This can be written in a nonlinear version of Hooke's law, as follows:

$$\sigma = E_0\varepsilon + E_1\varepsilon^2 + E_2\varepsilon^3 + \dots \tag{1.64}$$

When translated to macroscopic deformation, (1.64) implies that the stiffness of the beam will depend on the deformation; thus, the equation of motion will not be linear anymore. This source of nonlinearity can appear at any type of resonator, as it is directly linked to the material properties. However, it is typically only visible in either carefully engineered or very stiff structures[17, 18] where the rest of nonlinearities are suppressed.

1.3.1.2 Geometric Nonlinearity

The second most common type of nonlinearity is that associated with the geometry and boundary conditions of the resonator itself. This type of nonlinearity is mostly seen in flexural (bending) modes[19–22]. It is particularly common in clamped-clamped beams and strings, where a deflection implies an effective elongation of the beam. That elongation can only be achieved by means of developing a longitudinal stress within the beam, which in turn causes the frequency to change. As this developed tension is proportional to the amplitude of vibration, that is the origin of the nonlinearity that will be discussed in detail in Sect. 2.2.2.

1.3.1.3 Actuation Nonlinearity

The next source of nonlinearity to be considered here is the one emerging from a nonlinear dependence on the displacement of the actuation force[19, 20, 23, 24]. It is possible to understand it from the point of view of an actuation force that modifies the potential well of the harmonic oscillator. This example is found quite commonly in MEMS/NEMS resonators as it is inherent to electrostatic actuation, one of the most used actuation techniques for MEMS/NEMS. Electrostatic actuation-induced nonlinearity is discussed in detail in Sect. 4.2.1.3.

1.3.1.4 Detection Nonlinearity

After considering the nonlinearity in the actuation, it is only logical to follow by considering the nonlinearity in the detection. This is a very typical case of nonlinearity, present in the majority of experiments performed in MEMS/NEMS resonators. However, this does not affect the dynamics of the mechanical resonators, just the conversion between the *actual displacement* and the *measured voltage* in the macroscopic laboratory equipment. This type of nonlinearity becomes very important to determine the actual nonlinear parameters of the system[18, 22].

1.3.1.5 Nonlinear Damping

The final source of nonlinearity to be considered here is the one that affects the dissipation in the mechanical structure[14, 19, 25]. This is typically considered linear, i.e., constant no matter the displacement, but in reality there is no theoretical deterrent for the damping not to be nonlinear. In fact, in some extreme cases, this nonlinearity in the damping can be observed, and it can even dominate over the linear damping[26]. The origins of this nonlinear damping are still unclear and remain as an interesting field of fundamental research. In this case, the equation of motion (1.1) must be modified including terms of the type: $\eta z^2 \dot{z}$ or $\mu z \dot{z}$. An amplitude-dependent quality factor at large vibrational amplitudes can occur in resonators under tensile stress, as discussed in Sect. 3.3.3 in (3.58), which has been observed experimentally [27]. While this effect looks like nonlinear damping, the origin is a geometric effect based on linear damping.

1.3.1.6 Other Types of Nonlinearity

To finalize this section, it is important to remember that only a simplified case with one type of nonlinearity has been analyzed in detail. In reality there might be many nonlinear terms affecting the resonator dynamics, and thus more complicated analysis needs to be done. However, it can be shown that all the nonlinear effects can be analyzed in the first order using the nonlinear stiffness that has been shown here and the nonlinear damping. As stated before, for a more detailed analysis, the reader is invited to visit [14].

1.4 Parametric Amplification

Vibrations of a resonator can not only be induced by an external force $F(t)$ but also by periodic variations of the resonator's parameters. This effect is related to the well-known "pumping" of a playground swing. This effect can also occur in nanomechanical resonators. On the one hand, it can be externally induced by means of an appropriate transduction design. On the other hand, it can happen intrinsically when a nanomechanical resonator is driven in the geometric nonlinear regime, as it will be discussed in Sect. 2.7.3.

It's possible that any of the parameters of a linear damped resonator (mass, damping coefficient, and spring constant) can vary with time. In nanomechanical resonators however, it's typically the spring that undergoes periodic variations. For the sake of simplicity, the parametric resonance phenomenon is discussed here for an undamped linear system. The equation of motion of a linear undamped oscillator (1.4) with a time-dependent spring constant $k(t)$, and hence a time-dependent eigenfrequency $\omega_0^2(t) = k(t)/m$, can be written as

$$\ddot{z} + \omega_0^2(t)z = 0. \tag{1.65}$$

Let's assume a small parametric variation of the spring constant and hence the time-dependent eigenfrequency can be set to

$$\omega_0(t) = \omega_0 \left(1 + \alpha_p \cos \omega t\right) \quad (\alpha_p \ll 1). \tag{1.66}$$

Dismissing the α_p^2-term, (1.65) becomes

$$\ddot{z} + \omega_0^2 \left[1 + 2\alpha_p \cos (\omega t)\right] z = 0. \tag{1.67}$$

This equation of motion can slightly be reorganized into the following form:

$$\ddot{z} + \omega_0^2 z = -2\alpha_p \omega_0^2 z \cos (\omega t). \tag{1.68}$$

This equation of motion can be interpreted as a driven linear resonator. Importantly, the driving force on the right side of the equation is a linear function of the vibrational amplitude z, which has interesting implications. First, if the oscillator is at rest ($z(t) = 0$), the driving force is zero as well. Hence, in order for this parametric driving force to act, it's required that the system is oscillating already. This is distinctly different compared to a regularly driven linear resonator. Second, in the case that the system is oscillating at its eigenfrequency $z(t) = z_0 \cos \omega_0 t$, the driving force becomes proportional to

$$\cos \omega_0 t \cos \omega t = \frac{1}{2} \cos (\omega - \omega_0)t + \frac{1}{2} \cos (\omega + \omega_0)t. \tag{1.69}$$

It's clear that the parametric driving force has a component that matches the eigenfrequency when $\omega = 2\omega_0$. In that case, if the parametric driving force is stronger than the effective damping force, the vibration builds up exponentially creating the parametric resonance. This intuitive understanding allows now the definition of an ansatz for the solution of (1.67), which is a steady-state oscillation at ω_0 multiplied by an exponential growth function:

$$z(t) = z_0 \cos (\omega_0 t)e^{\mu t}, \tag{1.70}$$

with the exponential growth rate μ. Let's assume that $\mu \ll \omega_0$. Inserting (1.70) into (1.67) and dismissing the μ^2-term yields

$$2e^{\mu t} \omega_0 z_0 \left[\alpha_p \omega_0 \cos (\omega_0 t) \cos (2\omega_0 t) - \mu \sin (\omega_0 t)\right] = 0 \tag{1.71}$$

Let's only consider the time-independent solution; then, the parametric growth rate is

$$\mu = \alpha_p \omega_0. \tag{1.72}$$

As mentioned above, exponential parametric resonance only occurs if the growth rate is larger than the damping rate of the system. From the solution of a free damped system (1.9), the exponential damping term is known. It is straightforward to define a solution for a damped parametrically pumped system:

$$z(t) = z_0 \cos{(\omega_0 t)} e^{\alpha_p \omega_0 t} e^{-\zeta \omega_0 t} = z_0 \cos{(\omega_0 t)} e^{(\alpha_p - \zeta)\omega_0 t}. \tag{1.73}$$

It is now obvious that the threshold for parametric resonance occurs if the parametric growth rate is larger than the damping ratio:

$$\alpha_p > \zeta. \tag{1.74}$$

In this case the system vibrational amplitude increases exponentially until it gets limited by system nonlinearities. Hence, a parametrically pumped nanomechanical system can strongly amplify a small vibrational signal and effectively improve the mechanical FWHM, which is a promising technique, e.g., for nanomechanical force sensing. Parametric amplification has been implemented, among others, with nanomechanical beam resonators [28, 29], carbon nanotube resonators [30], and silicon nitride drum resonators [31]. If the parametric pumping stays below the threshold for parametric resonance, it's possible to squeeze the thermomechanical noise [32, 33]. Finally, it's been demonstrated that it is possible to create a nanomechanical parametric feedback oscillator by feeding back the detection signal to pump the parametric resonance [34]. This is in contrast to a conventional self-sustaining oscillator, as discussed in Sect. 5.2.2.3, where the feedback signal is used to drive the nanomechanical resonator directly.

References

1. Bao, M. (2005). *Analysis and design principles of MEMS devices*. Elsevier.
2. Albrecht, T.R., Grütter, P., Horne, D., & Rugar, D. (1991). Frequency modulation detection using high-Q cantilevers for enhanced force microscope sensitivity. *Journal of Applied Physics, 69*(2), 668.
3. Gil-Santos, E., Ramos, D., Jana, A., Calleja, M., Raman, A., & Tamayo, J. (2009). Mass sensing based on deterministic and stochastic responses of elastically coupled nanocantilevers. *Nano Letters, 9*(12), 4122–4127.
4. Stassi, S., De Laurentis, G., Chakraborty, D., Bejtka, K., Chiodoni, A., Sader, J. E., & Ricciardi, C. (2019). Large-scale parallelization of nanomechanical mass spectrometry with weakly-coupled resonators. *Nature Communications, 10*(3647), 3647.
5. Spletzer, M., Raman, A., Sumali, H., & Sullivan, J. P. (2008) Highly sensitive mass detection and identification using vibration localization in coupled microcantilever arrays. *Applied Physics Letters, 92*(11), 2006–2009.

6. Thiruvenkatanathan, P., Yan, J., Woodhouse, J., & Seshia, A. A. (2009). Enhancing parametric sensitivity in electrically coupled MEMS resonators. *Journal of Microelectromechanical Systems, 18*(5), 1077–1086.
7. Pourkamali, S., & Ayazi, F. (2005). Electrically coupled MEMS bandpass filters: Part I: With coupling element. *Sensors and Actuators, A: Physical, 122*(2), 307–316.
8. Aspelmeyer, M., Kippenberg, T. J., & Marquardt, F. (2014). Cavity optomechanics. *Reviews of Modern Physics, 86*(4), 1391–1452.
9. Novotny, L. (2010). Strong coupling, energy splitting, and level crossings: A classical perspective. *American Journal of Physics, 78*(11), 1199.
10. Garrido Alzar, C. L., Martinez, M. A. G., & Nussenzveig, P. (2002). Classical analog of electromagnetically induced transparency. *American Journal of Physics, 70*(1), 37–41.
11. Bagci, T., Simonsen, A., Schmid, S., Villanueva, L. G., Zeuthen, E., Appel, J., Taylor, J. M., Sørensen, A., Usami, K., Schliesser, A., & Polzik, E. S. (2014). Optical detection of radio waves through a nanomechanical transducer. *Nature, 507*(7490), 81–85.
12. Safavi-Naeini, A. H., Mayer Alegre, T. P., Chan, J., Eichenfield, M., Winger, M., Lin, Q., Hill, J. T., Chang, D. E., & Painter, O. (2011). Electromagnetically induced transparency and slow light with optomechanics. *Nature, 472*(7341), 69–73.
13. Weis, S., Riviére, R., Deléglise, S., Gavartin, E., Arcizet, O., Schliesser, A., & Kippenberg, T. J. (2010). Optomechanically induced transparency. *Science, 330*(6010), 1520–1523.
14. Lifshitz, R., & Cross, M. C. (2008). *Nonlinear dynamics of nanomechanical and micromechanical resonators* (Vol. 1), book section 1. Wiley-VCH, Weinheim.
15. Nayfeh, A. H., & Mook, D. T. (1979). *Nonlinear oscillations*. Pure and applied mathematics. Wiley.
16. Gere, J.M., & Goodno, B. J. (2013). *Mechanics of materials* (8th Ed.). Cengage Learning.
17. Kaajakari, V., Mattila, T., Oja, A., & Seppa, H. (2004). Nonlinear limits for single-crystal silicon microresonators. *Journal of Microelectromechanical Systems, 13*(5), 715–724.
18. Matheny, M. H., Villanueva, L. G., Karabalin, R. B., Sader, J. E., & Roukes, M. L. (2013). Nonlinear Mode-Coupling in Nanomechanical Systems. *Nano Letters, 13*(4), 1622–1626.
19. Kozinsky, I., Postma, H. W. C., Bargatin, I., & Roukes, M. L. (2006). Tuning nonlinearity, dynamic range, and frequency of nanomechanical resonators. *Applied Physics Letters, 88*(25), 253101.
20. Kacem, N., Hentz, S., Pinto, D., Reig, B., & Nguyen, V. (2009). Nonlinear dynamics of nanomechanical beam resonators: Improving the performance of NEMS-based sensors. *Nanotechnology, 20*(27), 275501.
21. Kacem, N., Arcamone, J., Perez-Murano, F., & Hentz, S. (2010). Dynamic range enhancement of nonlinear nanomechanical resonant cantilevers for highly sensitive NEMS gas/mass sensor applications. *Journal of Micromechanics and Microengineering, 20*(4), 45023.
22. Villanueva, L. G., Karabalin, R. B., Matheny, M. H., Chi, D., Sader, J. E., & Roukes, M. L. (2013). Nonlinearity in nanomechanical cantilevers. *Physical Review B, 87*(2), 24304.
23. San Paulo, A., & Garcia, R. (2001). Tip-surface forces, amplitude, and energy dissipation in amplitude-modulation (tapping mode) force microscopy. *Physical Review B, 64*(19), 193411.
24. Rhoads, J. F., Shaw, S. W., & Turner, K. L. (2010). Nonlinear dynamics and its applications in micro- and nanoresonators. *Journal of Dynamic Systems Measurement and Control-Transactions of the ASME, 132*(3), 34001 (2010)
25. Stav Zaitsev, Oleg Shtempluck, Eyal Buks, and Oded Gottlieb (2012). Nonlinear damping in a micromechanical oscillator. *Nonlinear Dynamics, 67*(1), 859–883.
26. Eichler, A., Moser, J., Chaste, J., Zdrojek, M., Wilson-Rae, I., & Bachtold, A. (2011). Nonlinear damping in mechanical resonators made from carbon nanotubes and graphene. *Nature Nanotechnology, 6*(6), 339–342 (2011)
27. Catalini, L., Rossi, M., Langman, E. C., & Schliesser, A. (2021). Modeling and observation of nonlinear damping in dissipation-diluted nanomechanical resonators. *Physical Review Letters, 126*(17), 174101.
28. Karabalin, R. B., Feng, X. L., Roukes, M. L. (2009). Parametric nanomechanical amplification at very high frequency. *Nano Letters, 9*(9), 3116–3123 (2009). PMID: 19736969.

29. Karabalin, R. B., Masmanidis, S. C., Roukes, M. L. (2010). Efficient parametric amplification in high and very high frequency piezoelectric nanoelectromechanical systems. *Applied Physics Letters, 97*(18), 183101.

30. Eichler, A., Chaste, J., Moser, J., & Bachtold, A. (2011). Parametric amplification and self-oscillation in a nanotube mechanical resonator. *Nano Letters, 11*(7), 2699–2703. PMID: 21615135.

31. Wu, S., Sheng, J., Zhang, X., Wu, Y., & Wu, H. (2018). Parametric excitation of a sin membrane via piezoelectricity. *AIP Advances, 8*(1), 015209.

32. Suh, J., LaHaye, M. D., Echternach, P. M., Schwab, K. C., & Roukes, M. L. (2010). Parametric amplification and back-action noise squeezing by a qubit-coupled nanoresonator. *Nano Letters, 10*(10), 3990–3994. PMID: 20843059.

33. Miller, J. M., Shin, D. D., Kwon, H. K., Shaw, S. W., & Kenny, T. W. (2020). Spectral narrowing of parametrically pumped thermomechanical noise. *Applied Physics Letters, 117*(3), 033504.

34. Villanueva, L. G., Karabalin, R. B., Matheny, M. H., Kenig, E., Cross, M. C., & Roukes, M. L. (2011). A nanoscale parametric feedback oscillator. *Nano Letters*, 11(11):5054–5059. PMID: 22007833.

Chapter 2
Continuum Mechanical Resonators

In this chapter, analytical models are introduced to describe the normal modes of continuum mechanical structures typically used as micro- and nanomechanical resonators. The chapter starts with the introduction of the *Rayleigh-Ritz method* in Sect. 2.1. This method presents a general method to approximate the eigenfrequency of a specific mechanical boundary value problem. Afterward, bending vibrations of "one-dimensional" resonators, such as *beams* and *strings*, are introduced in Sect. 2.2. A string is a special case of a doubly clamped beam that is under tensile stress, such that the effect of the tensile stress is dominating over the beam's bending stiffness (flexural rigidity). In other words, the difference between beams and strings lies in the way potential energy is stored. While in beams the potential energy is stored in the elastic bending, in strings it is stored in the work done against the strong tensile stress during deflection. Bending vibrations of "two-dimensional" plates are introduced in Sect. 2.3. Here the distinction is made between pure stress-free *plates* and pre-stressed *membranes/drumheads*, similar to the distinction between beams and strings in the one-dimensional case. In Sect. 2.4 "one-dimensional" bulk modes are described. In Sect. 2.5 torsional resonances in slender beams are discussed.

In Sect. 2.6 on page 60, it is explained how the effective parameters of a specific continuum mechanical resonator are calculated in order to reduce it to a lumped-element model system discussed in the previous Chap. 1. Finally, in Sect. 2.7 on page 62, geometric nonlinearities are discussed, which commonly occur in doubly clamped beams/strings and fixed plates/membranes.

2.1 Rayleigh-Ritz Method

The classical way to find the eigenmodes of a continuum mechanical structures is by deriving the differential equation of motion, which can be obtained by equalizing all

S. Schmid et al., *Fundamentals of Nanomechanical Resonators*,
https://doi.org/10.1007/978-3-031-29628-4_2

the forces (inertial and elastic) acting on an infinitesimally small piece of structure (Newton's third law). In this way it is readily possible to derive exact solutions for many simple and idealized continuum mechanical structures, such as beams, strings, or membranes. But for some structures, such as plates, it can be very hard to find the solution of the resulting differential equations. In that case the eigenmodes can be approximated by methods based on energy assumptions.

Rayleigh's method is a useful tool to obtain a good approximation for the fundamental eigenfrequency of a conservative system (a system without loss of energy). The method is based on the complete commute of the energy between its kinetic and potential state (equipartition theorem), as already applied previously in Sect. 1.1.1. Hence, we can assume that the maximal kinetic energy ($W_{kin,max}$) must be equal to the maximal potential energy ($W_{pot,max}$):

$$W_{kin,max} = W_{pot,max}. \tag{2.1}$$

All the potential and kinetic energies of a few common ideal structures introduced in this chapter are listed in Table 2.1. It is obvious that in order to calculate the energies, the structure-specific mode shape $U(x, y, t)$ is required. It is possible to separate the spatial modal shape from the temporal variations. Separating the variables, the mode shape function of a specific normal mode can be written in the form

$$U(x, y, t) = \phi(x, y)u(t) = \phi(x, y)u_0 \cos(\omega_0 t) \tag{2.2}$$

where the spatial mode shape function $\phi(x, y)$ gets modulated by the temporal sinusoidal function $U_0 \cos(\omega_0 t)$ with amplitude u_0 and angular velocity ω_0, which is the eigenfrequency of the system.[1] The separation of variables is done likewise with polar coordinates.

Since the maximal value of the temporal function is unity ($\max\{\cos(\omega_0 t)\} = 1$) and the potential energy is proportional to the vibrational amplitude squared

$$W_{pot} \propto (U(x, y, t))^2, \tag{2.3}$$

the potential energy maximum is readily given by

$$W_{pot,max} = \max\{W_{pot}(U(x, y, t))\} = u_0^2 W_{pot}(\phi(x, y)). \tag{2.4}$$

Similarly, because for the kinetic energy it is essential that

$$W_{kin} \propto \left(\frac{\partial U(x, y, t)}{\partial t}\right)^2, \tag{2.5}$$

---

[1] For convenience, the term "frequency" is subsequently used in place for the actual correct term "angular velocity."

Table 2.1 Strain and kinetic energies of various continuum mechanical structures with E: Young's modulus, I_y: geometrical moment of inertia, h: structure thickness, w: beam width, $A = hw$: cross section, ρ: mass density, G shear modulus, I_p: polar moment of inertia, v: Poisson's ratio, $D_P = Eh^3/(12(1 - v^2))$: flexural rigidity of plate, σ: tensile stress[1–4]

Member	Potential energy (W_{pot})	Kinetic energy (W_{kin})
Beam of length L in bending vibration	$\frac{1}{2}EI_y \int_0^L \left(\frac{\partial^2 U}{\partial x^2}\right)^2 dx$	$\frac{1}{2}A\rho \int_0^L \left(\frac{\partial U}{\partial t}\right)^2 dx$
String of length L in bending vibration	$\frac{1}{2}\sigma A \int_0^L \left(\frac{\partial U}{\partial x}\right)^2 dx$	
Beam or **String** of length L in bending vibration	Energy stored in longitudinal extension $\frac{1}{8}EA \int_0^L \left(\frac{\partial U}{\partial x}\right)^4 dx$	
Rod of length L in longitudinal vibration	$\frac{1}{2}EA \int_0^L \left(\frac{\partial U}{\partial x}\right)^2 dx$	
Rod of length L in torsional vibration	$\frac{1}{2}GI_p \int_0^L \left(\frac{\partial U}{\partial x}\right)^2 dx$	$\frac{1}{2}I_p\rho \int_0^L \left(\frac{\partial U}{\partial t}\right)^2 dx$
Rectangular plate of size S in bending vibration	$\frac{D_P}{2} \iint_S \left\{ \left(\frac{\partial^2 U}{\partial x^2} + \frac{\partial^2 U}{\partial y^2}\right)^2 + 2(1-v)\left[\left(\frac{\partial^2 U}{\partial x \partial y}\right)^2 - \frac{\partial^2 U}{\partial x^2}\frac{\partial^2 U}{\partial y^2}\right] \right\} dxdy$	$\frac{1}{2}\rho h \iint_S \left(\frac{\partial U}{\partial t}\right)^2 dxdy$
Circular plate of radius R in center symmetrical bending vibration	$\pi D_P \int_0^R \left\{ \left(\frac{\partial^2 U}{\partial r^2} + \frac{1}{r}\frac{\partial U}{\partial r}\right)^2 - 2(1-v)\frac{\partial^2 U}{\partial r^2}\frac{1}{r}\frac{\partial U}{\partial r} \right\} rdr$	$\pi\rho h \int_0^R r\left(\frac{\partial U}{\partial t}\right)^2 dr$
Rectangular membrane of size S in bending vibration	$\frac{1}{2}\sigma h \iint_S \left\{ \left(\frac{\partial U}{\partial x}\right)^2 + \left(\frac{\partial U}{\partial y}\right)^2 \right\} dxdy$	$\frac{1}{2}\rho h \iint_S \left(\frac{\partial U}{\partial t}\right)^2 dxdy$
Circular membrane of radius R in bending vibration	$\sigma h\pi \int_0^R \left(\frac{\partial U}{\partial r}\right)^2 rdr$	$\pi\rho h \int_0^R r\left(\frac{\partial U}{\partial t}\right)^2 dr$

and with the Ansatz (2.2), the maximal kinetic energy readily can be written as

$$W_{kin,max} = \max\{W_{kin}(U(x, y, t)) = \omega_0^2 u_0^2 W_{kin}(\phi(x, y)). \qquad (2.6)$$

Substituting (2.4) and (2.6) in (2.1), it is now possible to calculate the eigenfrequency ω_0 of a specific eigenmode:

$$\omega_0^2 = \frac{W_{pot}(\phi(x, y))}{W_{kin}(\phi(x, y))} \qquad (2.7)$$

for a suitable mode shape function $\phi(x, y)$. Rayleigh's method yields the exact eigenfrequency if the exact mode shape is known. But typically, the correct mode shape function is unknown, and a suitable approximation has to be assumed that satisfies all the boundary conditions. A common approach is to use the displacement function of the respective static deflection of the structure of interest due to a uniform load. An assumed approximate mode shape is always stiffer than the correct function, and the resulting eigenfrequency obtained by Rayleigh's method is always slightly higher than the exact eigenfrequency.

Rayleigh's method gives a decent approximation for the fundamental eigenfrequency. But in order to obtain more accurate solutions, also of higher modes, Rayleigh's method is often combined with the *Ritz method*. This method is a so-called variational method used to approximate various mechanical boundary value problems. The Ritz method is based on minimizing the total potential energy in a mechanical system. The minimum energy is found by varying free variables of a test solution that obeys all boundary conditions. The combined method to approximate the eigenfrequencies of specific mechanical boundary value problems is often called the *Rayleigh–Ritz method*. According to the Ritz method, n free variables c_i are introduced to the spatial mode shape function $\phi(x, y)$[1]:

$$\phi(x, y) = \sum_{i=1}^{n} c_i \phi_i(x, y). \tag{2.8}$$

The set of variables c_i that minimize the total energy in the structure resulting in a minimum eigenfrequency can be found by minimizing the so-called Rayleigh quotient (2.7):

$$\frac{\partial}{\partial c_i} \omega_0^2 = \frac{\partial}{\partial c_i} \left\{ \frac{W_{pot}(\phi(x, y))}{W_{kin}(\phi(x, y))} \right\} = 0 \quad (i = 1, 2, 3, \ldots, n) \tag{2.9}$$

which by applying the quotient derivation rule and with (2.7) becomes

$$\frac{\partial}{\partial c_i} \{ W_{pot}(\phi(x, y)) - \omega_0^2 W_{kin}(\phi(x, y)) \} = 0 \quad (i = 1, 2, 3, \ldots, n). \tag{2.10}$$

Substituting (2.8) into (2.10) yields a homogeneous linear system of equations of n variables, which can be written as

$$\mathbf{M}_n(\omega_0)\mathbf{c}_i = 0 \tag{2.11}$$

with the square matrix $\mathbf{M}_n(\omega)$ of order n multiplied by the c_i values of the vector $\mathbf{c}_i$. In order to get the nontrivial solution, the determinant of this system has to be equal to zero

$$\det(\mathbf{M}_n(\omega_0)) = 0. \tag{2.12}$$

This results in the *frequency* or *characteristic equation* yielding the corresponding n eigenfrequencies ω_i. It is clear from (2.10) that the Rayleigh-Ritz method for $n = 1$ reduces to the Rayleigh method (2.7).

In this chapter, the eigenfrequencies of a few exemplary structures, such as beams, strings, and rectangular membranes, are derived exactly. For all other structures, the fundamental eigenfrequency is approximated by Rayleigh's method, yielding the basic terms of the particular eigenfrequencies. Approximations for the higher modes, which typically were derived by the Rayleigh-Ritz method, are taken from mechanics textbooks.

2.2 One-Dimensional Bending Vibrations

One-dimensional bending vibrations of beams and strings are among the most common nanomechanical structures. By some definitions, they are the only actual nanomechanical resonators with at least two dimensions below the size of 1 μm. A few typical examples are shown in Fig. 2.1.

2.2.1 Free Bending Vibration of Beams

The eigenmode problem of bending beams is one of the prominent cases that can be solved exactly. The model is based on the beam bending theory finalized in the eighteenth century by the Swiss mathematicians Leonard Euler and Daniel Bernoulli, hence the name *Euler-Bernoulli beam theory*. The bending behavior of beams can be modeled under the of assumptions that the beam is slender ($L/h > 10$) (see Fig. 2.2) and that the rotational inertia and the shear deformation can be neglected. The equation of motion of a thin beam can be derived by means of the equilibrium of forces for an infinitesimal piece of beam. Assuming a linear elastic material and small deflections $U(x, t)$, the equation of motion of a thin beam (Euler-Bernoulli beam) is given by [9, 10]

$$\rho A \frac{\partial^2 U(x, t)}{\partial t^2} + E I_y \frac{\partial^4 U(x, t)}{\partial x^4} = 0 \tag{2.13}$$

where ρ is the mass density, A is the cross-sectional area, E is the Young's modulus, and I_y is the geometric moment of inertia with respect to the y-axis

$$I_y = \iint_V z^2 dA. \tag{2.14}$$

The two most common cross sections are rectangular and circular. The respective geometrical moments of inertia for a rectangular cross section, as depicted in Fig. 2.2, are given by

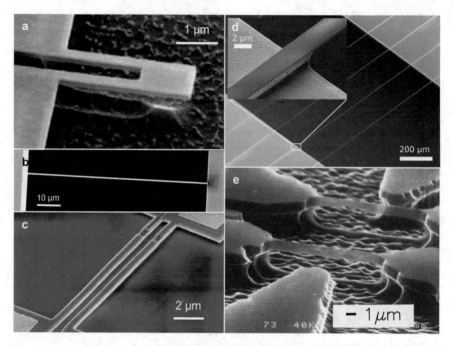

Fig. 2.1 Examples of flexural one-dimensional nanomechanical resonators. (**a**) Silicon nitride nanocantilever for gravimetric gas chromatography (Reprinted with permission from [5]. Copyright 2010 American Chemical Society.) (**b**) Silicon nitride nanostring for airborne nanoparticle detection [6]. (**c**) Aluminum nitride beam resonators. (**d**) High-Q silicon nitride string resonators. (Reprinted with permission from [7]. Copyright 2011 by the American Physical Society.) (**e**) Silicon nanobridges. (Reprinted from [8], with permission from AIP Publishing)

Fig. 2.2 Schematic drawing of a single-clamped beam

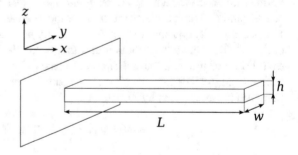

$$I_y = \frac{Ah^2}{12} \tag{2.15}$$

and for a circular cross section with radius R

$$I_y = \frac{\pi R^4}{4}. \tag{2.16}$$

The solution to this differential equation is a superposition of normal modes that can be separated into a position-dependent and a time-dependent term via a separation of variables:

$$U(x,t) = \sum_{n=1}^{\infty} \phi_n(x) u_n(t) = \sum_{n=1}^{\infty} \phi_n(x) u_{0,n} \cos(\omega_n t), \qquad (2.17)$$

where ω_n is the eigenfrequency and n denotes the modal number. A general solution to the mode shape function of the beam $\phi_n(x)$ can be written in the form

$$\phi_n(x) = a_n \cos \beta_n x + b_n \sin \beta_n x + c_n \cosh \beta_n x + d_n \sinh \beta_n x \qquad (2.18)$$

with the *wavenumber* β_n. The first two terms with the trigonometric functions represent the standing waves in the beam center, while the last two hyperbolic terms represent the influence of the clamping. From this equation it is clear that a beam will vibrate in certain vibrational modes each with a distinct spatial shape.

By insertion of Eq. (2.18) into (2.13), the differential equation can be rewritten as

$$- \rho A \omega^2 U(x,t) + EI_y \beta_n^4 U(x,t) = 0 \qquad (2.19)$$

which results in the *dispersion relationship* which yields the eigenfrequency ω_n as a function of the wavenumber

$$\omega_n = \beta_n^2 \sqrt{\frac{EI_y}{\rho A}}. \qquad (2.20)$$

The dispersion relationship can be written as

$$\omega_n = \beta_n^2 c_E \sqrt{\frac{I_y}{A}}. \qquad (2.21)$$

with the wave velocity c_E in the elastic beam

$$c_E = \sqrt{\frac{E}{\rho}}. \qquad (2.22)$$

Assuming a rectangular cross section with beam thickness h the geometric moment of inertia is given by (2.15). By defining a flexural rigidity of a square beam

$$DE = \frac{Eh^3}{12},$$
(2.23)

the eigenfrequency of a beam can be written as

$$\omega_n = \beta_n^2 \sqrt{\frac{D_E}{\rho h}}.$$
(2.24)

The Euler-Bernoulli beam theory assumes a thin and long beam. If the beam width to height ratio becomes larger $w/h > 5$, the flexural rigidity of a Euler-Bernoulli beam has to be replaced by the flexural rigidity of a plate:

$$D_P = \frac{Eh^3}{12(1 - \nu^2)}$$
(2.25)

where ν is the Poisson's ratio to account for the suppression of the in-plane dilatation accompanying axial strain which makes a plate stiffer than a beam [1, 3].

In the next step, the wavenumber of specific eigenmodes of a beam is derived. This is done by finding the unknown coefficients a_n, b_n, c_n, d_n in (2.18) by means of the specific boundary conditions of the beam. We take into account two specific cases: (1) singly clamped beams, also called cantilevers, and (2) doubly clamped beams, also called bridges.

2.2.1.1   Cantilevers

Cantilevers are fixed at one end, while the free end is curvature-free and does not experience a momentum. The boundary conditions of a cantilever are hence described by

$$\phi_n(0) = \frac{\partial}{\partial x}\phi_n(0) = \frac{\partial^2}{\partial x^2}\phi_n(L) = \frac{\partial^3}{\partial x^3}\phi_n(L) = 0.$$
(2.26)

These four boundary conditions create a system of linear equations of fourth order, which can be written as

$$\begin{bmatrix} 1 & 0 & 1 & 0 \\ 0 & 1 & 0 & 1 \\ -\cos(\beta_n L) & -\sin(\beta_n L) & \cosh(\beta_n L) & \sinh(\beta_n L) \\ \sin(\beta_n L) & -\cos(\beta_n L) & \sinh(\beta_n L) & \cosh(\beta_n L) \end{bmatrix} \begin{Bmatrix} a \\ b \\ c \\ d \end{Bmatrix} = \begin{Bmatrix} 0 \\ 0 \\ 0 \\ 0 \end{Bmatrix}.$$
(2.27)

A nontrivial solution exists for this homogeneous system if the determinant is zero, that is,

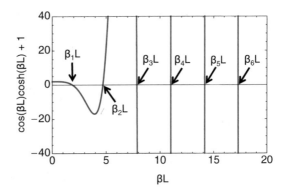

Fig. 2.3 Depiction of the roots of the frequency equation (2.28) for a cantilever

$$\cos(\beta_n L)\cosh(\beta_n L) + 1 = 0 \qquad (2.28)$$

which is the frequency equation. The equation is plotted in Fig. 2.3. From this equation has discrete solutions for specific wavenumbers β_n, which correspond to the specific eigenfrequencies of the cantilever. This transcendent equation can be solved numerically for the lower-order modes. The cosine is a periodic function, while the hyperbolic cosine is exponentially increasing with increasing $\beta_n L$. The number of roots of (2.28) is hence corresponding to the number of periods of $\cos(\beta_n L)$. For higher eigenvalues (2.28) simplifies to

$$\cos(\beta_n L) \approx 0 \;\; \forall \; n \geq 3 \qquad (2.29)$$

and $\beta_n L \approx (2n - 1)\pi/2$. In conclusion, the roots of the frequency equation of a cantilever beam are

$$\lambda_n = \beta_n L = 1.8751, 4.6941, 7.8548, (2n - 1)\pi/2. \qquad (2.30)$$

We can now write the eigenfrequency of a cantilever as

$$\omega_n = \frac{\lambda_n^2}{L^2}\sqrt{\frac{EI_y}{\rho A}}. \qquad (2.31)$$

The mode shape function $\phi_n(x)$ of a cantilever can be obtained from the boundary conditions (2.26). From the first two boundary conditions, we obtain:

$$\phi_n(0) = 0 : a_n + c_n = 0 \qquad (2.32)$$

$$\frac{\partial \phi_n(0)}{\partial x} = 0 : b_n + d_n = 0 \qquad (2.33)$$

and (2.18) reduces to

$$\phi_n(x) = a_n(\cos \beta_n x - \cosh \beta_n x) + b_n(\sin \beta_n x - \sinh \beta_n x). \tag{2.34}$$

With the third boundary condition, we obtain the ratio of the coefficients a_n and b_n:

$$\frac{\partial^2}{\partial x^2}\phi_n(L) = 0 : \quad \frac{b_n}{a_n} = -\frac{\cos(\beta_n L) + \cosh(\beta_n L)}{\sin(\beta_n L) + \sinh(\beta_n L)} \tag{2.35}$$

and (2.18) becomes

$$\phi_n(x) = a_n[\cos \beta_n x - \cosh \beta_n x - \frac{\cos(\beta_n L) + \cosh(\beta_n L)}{\sin(\beta_n L) + \sinh(\beta_n L)}(\sin \beta_n x - \sinh \beta_n x)]. \tag{2.36}$$

The first four mode shapes of a cantilever are shown in Fig. 2.4. It can be seen from the figure that certain areas of the cantilever have a large vibrational amplitude, whereas other areas (near the nodal points) are moving with low amplitude. The number of nodal points increases with increasing mode number.

The mode shape function is used in a normalized form with $\phi_n(L) = 1$. In a normalized form, which is satisfied for

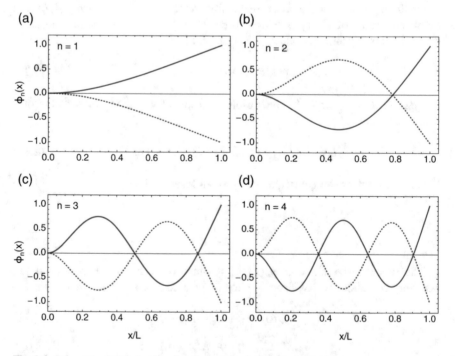

Fig. 2.4 Schematics of the first four bending modes (**a–d**) of a cantilever

$$\phi_n(x) = \frac{1}{2}[\cos \beta_n x - \cosh \beta_n x - \frac{\cos(\beta_n L) + \cosh(\beta_n L)}{\sin(\beta_n L) + \sinh(\beta_n L)}(\sin \beta_n x - \sinh \beta_n x)].$$

$$(2.37)$$

2.2.1.2 Bridges

Bridges are fixed at both ends, and the boundary conditions are hence described by

$$\phi_n(0) = \phi_n(L) = \frac{\partial}{\partial x}\phi_n(0) = \frac{\partial}{\partial x}\phi_n(L) = 0. \tag{2.38}$$

These four boundary conditions create a system of equations of fourth order, which can be written as

$$\begin{bmatrix} 1 & 0 & 1 & 0 \\ 0 & 1 & 0 & 1 \\ \sin(\beta_n L) & -\cos(\beta_n L) & \sinh(\beta_n L) & \cosh(\beta_n L) \\ \cos(\beta_n L) & +\sin(\beta_n L) & \cosh(\beta_n L) & \sinh(\beta_n L) \end{bmatrix} \begin{Bmatrix} a \\ b \\ c \\ d \end{Bmatrix} = \begin{Bmatrix} 0 \\ 0 \\ 0 \\ 0 \end{Bmatrix}. \tag{2.39}$$

Setting the determinant zero, we obtain the frequency equation:

$$\cos(\beta_n L)\cosh(\beta_n L) - 1 = 0 \tag{2.40}$$

with the the solutions for $n = 1, 2, 3, n > 3$ are $\lambda_n = \beta_n L = 4.7300, 7.8532, 10.9956, (2n + 1)\pi/2$, respectively. The frequency equation for a bridge is plotted in Fig. 2.5.

The mode shape function of a bridge can be obtained from the boundary conditions (2.38). From the first two boundary conditions we obtain, equal to the case of cantilevers

$$\phi_n(0) = 0: \ a + c = 0 \tag{2.41}$$

Fig. 2.5 Plot of the roots of the frequency equation 2.40 for a bridge

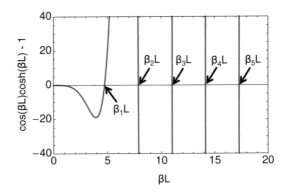

$$\frac{\partial \phi_n(0)}{\partial x} = 0 : \ b + d = 0 \tag{2.42}$$

and (2.18) reduces to

$$\phi_n(x) = a_n(\cos \beta_n x - \cosh \beta_n x) + b_n(\sin \beta_n x - \sinh \beta_n x). \tag{2.43}$$

With the third boundary condition, we obtain the ratio of the coefficients a_n and b_n:

$$\frac{\partial^2}{\partial x^2}\phi_n(L) = 0 : \ \frac{b_n}{a_n} = -\frac{\cos(\beta_n L) - \cosh(\beta_n L)}{\sin(\beta_n L) - \sinh(\beta_n L)} \tag{2.44}$$

and (2.18) becomes

$$\phi_n(x) = a_n[\cos \beta_n x - \cosh \beta_n x - \frac{\cos(\beta_n L) - \cosh(\beta_n L)}{\sin(\beta_n L) - \sinh(\beta_n L)}(\sin \beta_n x - \sinh \beta_n x)]. \tag{2.45}$$

The first four mode shapes of a bridge are shown in Fig. 2.6.

The normalized mode shape function for the peak of a bridge is approaching

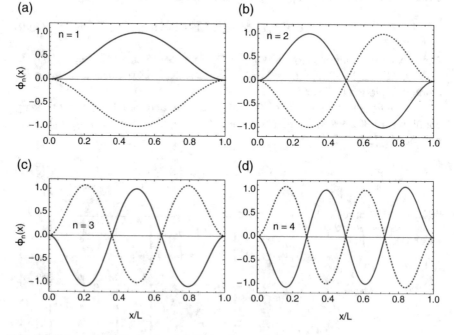

Fig. 2.6 Schematics of the first four bending modes (**a–d**) of a bridge

$$\phi_n(x) = \frac{1}{\sqrt{2}}[\cos\beta_n x - \cosh\beta_n x - \frac{\cos(\beta_n L) - \cosh(\beta_n L)}{\sin(\beta_n L) - \sinh(\beta_n L)}(\sin\beta_n x - \sinh\beta_n x)] \tag{2.46}$$

for $n > 3$. For lower mode numbers $n = 1, 2, 3, ...$, the normalization factors are $0.6297, 0.6626, 0.7112, \approx 1/\sqrt{2}$.

2.2.2 Free Bending Vibration of Beams Under Tensile Stress (Strings)

Thin films typically used in microfabrication tend to have a process-related tensile stress. Doubly clamped beams made of such thin films are therefore usually pre-stressed. A tensile stress σ increases the eigenfrequency and has to be taken into account by adding a term for the tensile force $N = \sigma A$ to (2.13). Applying Newton's third law, the free and undamped bending vibration for small amplitudes can then be described by

$$\rho A \frac{\partial^2 U(x, t)}{\partial t^2} + E I_y \frac{\partial^4 U(x, t)}{\partial x^4} - N \frac{\partial^2 U(x, t)}{\partial x^2} = 0 \tag{2.47}$$

This equation of motion can be solved with the ansatz (2.17) for the case of a simply supported (a support that allows a rotation of the clamping points) doubly clamped beam with the boundary conditions:

$$\phi_n(0) = \phi_n(L) = \frac{\partial^2}{\partial x^2}\phi_n(0) = \frac{\partial^2}{\partial x^2}\phi_n(L) = 0 \tag{2.48}$$

assuming a sinusoidal mode shape with wavenumber β_n

$$\phi_n(x) = \sin(\beta_n x). \tag{2.49}$$

Substituting (2.49) in the general ansatz (2.17), the equation of motion (2.47) readily yields the dispersion relationship which yields the eigenfrequency ω_n: as a function of β_n

$$\omega_n = \sqrt{\frac{\sigma}{\rho}\beta_n^2 + \frac{E I_y}{\rho A}\beta_n^4}. \tag{2.50}$$

This equation can be rearranged into the following form:

$$\omega_n = \beta_n^2 \sqrt{\frac{E I_y}{\rho A}}\sqrt{1 + \frac{\sigma A}{E I_y \beta_n^2}}. \tag{2.51}$$

It shows that the eigenfrequency is basically the eigenfrequency of an unstressed beam (2.20) multiplied by a term which increases with stress. Choosing a wavenumber that satisfies the boundary conditions (2.48)

$$\beta_n = \frac{n\pi}{L},$$ (2.52)

two extreme conditions can be defined. On the one hand, if

$$\frac{\sigma A}{E I_y} \frac{L^2}{(n\pi)^2} \ll 1$$ (2.53)

equation (2.51) reduces to (2.20) of an unstressed simply supported beam, as mentioned already. On the other hand, if

$$\frac{\sigma A}{E I_y} \frac{L^2}{(n\pi)^2} \gg 1$$ (2.54)

the flexural rigidity can be neglected and (2.51) reduces to

$$\omega_n = \frac{n\pi}{L} \sqrt{\frac{\sigma}{\rho}}$$ (2.55)

which is the eigenfrequency of a string, a pre-stressed threadlike elastic continuum without flexural rigidity. Compared to cantilevers or stress-free bridges, the resonance frequency of strings is not a function of the beam thickness. It is mainly defined by the length and the tensile stress. The eigenfrequency (2.55) can also be written in the following form:

$$\omega_n = \frac{n\pi}{L} c_\sigma$$ (2.56)

with the wave velocity inside a string of

$$c_\sigma = \sqrt{\frac{\sigma}{\rho}}.$$ (2.57)

The derivations above were done for a simply supported beam. These are valid boundary conditions for an ideal string with zero flexural rigidity. But a real micro- or nanomechanical stringlike resonator typically is clamped at both ends and possesses a non-negligible rigidity. As long as the string condition (2.54) is fulfilled, the beam rigidity can be neglected for the calculation of the eigenfrequency (2.55). However, the finite string rigidity will influence the mode shape, particularly at the clamping sites. The full set of wavenumbers required to model a suitable mode shape function of a string, on the basis of (2.18), can be found by solving the dispersion relation of the pre-stressed beam (2.50):

$$\beta_n^4 + \frac{\sigma A}{E I_y} \beta_n^2 - \omega_n^2 \frac{\rho A}{E I_y} = 0. \tag{2.58}$$

The four solutions of β_n are then

$$\beta_{n,1-4} = \pm \sqrt{\frac{1}{2} \frac{\sigma A}{E I_y} \left[-1 \pm \sqrt{1 + \underbrace{4\omega_n^2 \frac{\rho}{\sigma} \frac{E I_y}{\sigma A}}_{\ll 1}} \right]}. \tag{2.59}$$

Inserting the string eigenfrequency (2.55) shows that the last term in (2.59), according to the string condition (2.54), is a lot smaller than unity. Hence, the inner square root can be approximated by a Taylor series and (2.59) becomes:

$$\beta_{n,1-4} = \pm \sqrt{\frac{1}{2} \frac{\sigma A}{E I_y} \left[-1 \pm \left(1 + 2 \left(\frac{n\pi}{L} \right)^2 \frac{E I_y}{\sigma A} \right) \right]}. \tag{2.60}$$

In this form, the two positive solutions are apparent, namely,

$$\beta_{n,1} = \beta_\sigma = \frac{n\pi}{L}, \tag{2.61}$$

which is the already known wavenumber of a perfect string, and

$$\beta_{n,2} = \beta_E = \sqrt{\frac{\sigma A}{E I_y}} \tag{2.62}$$

which is the wave number related to the flexural stiffness of the string.

The mode shape of a "realistic" string resonator can now be approximated by using the string wavenumber β_σ for the trigonometric terms in (2.18) which represent the standing waves in the string. For the hyperbolic terms, representing the clamping conditions, the wavenumber β_E from the flexural stiffness of the string is used instead:

$$\phi_n(x) = a_n \cos \beta_\sigma x + b_n \sin \beta_\sigma x + c_n \cosh \beta_E x + d_n \sinh \beta_E x. \tag{2.63}$$

The first two boundary conditions for a doubly clamped beam (2.38) yield a simplified mode shape:

$$\phi_n(x) = a_n \left(\cos \beta_\sigma x - \cosh \beta_E x \right) + b_n \left(\sin \beta_\sigma x + \frac{\beta_\sigma}{\beta_E} \sinh \beta_E x \right). \tag{2.64}$$

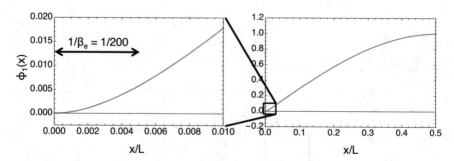

Fig. 2.7 String mode shape (2.67) for $\beta_E = 200$

The last two conditions result in $a_n \approx -\frac{\beta_\sigma}{\beta_E} b_n$ under the valid string assumption of $\beta_E \gg 1$ and the mode shape function can be expressed as

$$\phi_n(x) = \begin{cases} \phi_n(x) & , 0 \leq x \leq L/2 \\ (-1)^{n+1}\phi_n(L-x) & , L/2 < x \leq L \end{cases} \tag{2.65}$$

with

$$\phi_n(x) = \left(\sin(\beta_\sigma x) - \frac{\beta_\sigma}{\beta_E} \left[\cos(\beta_\sigma x) - \cosh(\beta_E x) + \sinh(\beta_E x) \right] \right), \tag{2.66}$$

which can be further simplified with the algebraic expressions $\sinh x = (e^x - e^{-x})/2$ and $\cosh x = (e^x + e^{-x})/2$ to the following form:

$$\phi_n(x) = \left(\sin(\beta_\sigma x) - \frac{\beta_\sigma}{\beta_E} \left[\cos(\beta_\sigma x) - e^{-\beta_E x} \right] \right). \tag{2.67}$$

This equation shows that β_E represents the inverse of the exponential edge-correction decay length which results from the flexural rigidity of the string. The string mode shape (2.67) and the edge effect resulting from the flexural rigidity of the string are visualized in Fig. 2.7.

2.3 Two-Dimensional Bending Vibrations

Two-dimensional bending resonators can be separated into *plates* and *membranes*, similar to the distinction between one-dimensional *beams* and *strings*, as introduced in Sect. 2.2 on page 35. The ideal case of a *plate* is reached if the mechanical behavior is dominated by the the bending stiffness (flexural rigidity) of the structure. The other ideal case of a *membrane* occurs if there is a tensile stress inside the structure that is dominating its behavior, and the flexural rigidity does not have

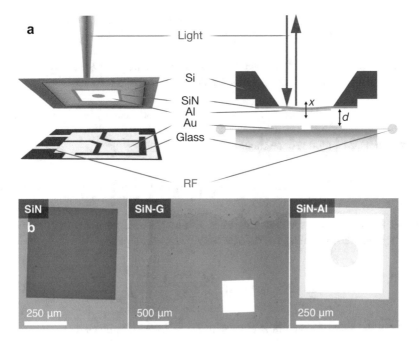

Fig. 2.8 Examples of two-dimensional flexural micromechanical resonators. (**a**) Aluminum-coated silicon nitride membrane resonator [9], (**b**) SiN membranes coated with a single graphene layer (SiN-G) and aluminum (SiN-Al) for optoelectromechanical coupling. (Reprinted from [11], with permission from AIP Publishing)

to be taken into account. They are interesting partly for two reasons. First, they can be fabricated from thin films with thicknesses reaching from a few tens of nanometers down to a few angstroms when, e.g., fabricated from single layers of graphene. Second, the large lateral dimensions allow for a strong optical or electrical interaction, which results in efficient transduction of the vibrational motion. The low mass combined with the strong external coupling is the reason that two-dimensional resonators have become key structures in many applications, most prominently in optomechanics. Typical examples of two-dimensional resonators are shown in Fig. 2.8.

All subsequently presented continuum mechanical models are based on the general assumptions of perfectly elastic, homogeneous, and isotropic materials.

2.3.1 Free Bending Vibration of Plates

2.3.1.1 Rectangular Plates

The equation of motion of a *rectangular plate* with a flexural rigidity D_P [see Eq. 2.25 on page 38] is given by the two-dimensional wave equation [1, 12]:

$$D_P \nabla^2 \nabla^2 U - \rho h \frac{\partial^2 U}{\partial t^2} = 0 \tag{2.68}$$

which, however, can only be solved for the case that the plate is simply supported with the common ansatz for a specific mode

$$U_{n,j}(x, y, t) = \phi_{n,j}(x, y) u_{0,n,j} \cos(\omega_{n,j} t) \tag{2.69}$$

with a sinusoidal mode shape function

$$\phi_{n,j}(x, y) = \sin\left(\frac{n\pi x}{L_x}\right) \sin\left(\frac{j\pi y}{L_y}\right) \tag{2.70}$$

where L_x and L_y are the side lengths in x- and y-direction. Inserting (2.69) into (2.68) yields the eigenfrequencies of a *simply supported* plate:

$$\omega_{n,j} = \pi^2 \sqrt{\frac{D_P}{\rho h}} \left(\frac{n^2}{L_x^2} + \frac{j^2}{L_y^2}\right) \tag{2.71}$$

which in the case of a quadratic plate with $L_x = L_y = L$ reduces to

$$\omega_{n,j} = \frac{\pi^2 (n^2 + j^2)}{L^2} \sqrt{\frac{D_P}{\rho h}}. \tag{2.72}$$

As presented, the problem of simply supported plates is straightforward to solve. But simply supported plates are basically impossible to realize on the micro- and nanoscale, and the more likely boundary conditions are, e.g., all four edges clamped or free. The eigenmode problem of these boundary conditions is very hard to solve. For this case, *Rayleigh's method* is a useful tool to obtain a good approximation for the fundamental eigenfrequency. The maximal kinetic and potential energies for a specific mode of a plate are given by (from Table 2.1)

$$W_{kin,max} = \frac{\omega_{n,j}^2}{2} \rho h \iint U^2 dx dy. \tag{2.73}$$

and

$$W_{pot,max} = \frac{1}{2} D_P \iint \left\{ (\nabla^2 U)^2 + 2(1 - v) \left[\left(\frac{\partial^2 U}{\partial x \partial y}\right)^2 - \frac{\partial^2 \phi}{\partial x^2} \frac{\partial^2 U}{\partial y^2} \right] \right\} dx dy. \tag{2.74}$$

For a rectangular plate, the following mode shape function describes the mode shape of the fundamental mode to a good approximation:

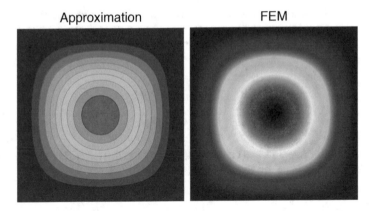

Fig. 2.9 Comparison of the plate displacement assumed for the analytical approximation and the actual displacement calculated by FEM

$$\phi_{1,1}(x, y) = c(x^2 - (L_x/2)^2)^2 (y^2 - (L_y/2)^2)^2 \tag{2.75}$$

and it satisfies the clamped boundary conditions:

$$\left. \frac{\partial \phi_{1,1}}{\partial x} = 0 \right|_{x = \pm L_x/2} \quad \& \quad \left. \frac{\partial \phi_{1,1}}{\partial y} = 0 \right|_{y = \pm L_y/2} . \tag{2.76}$$

A comparison of this mode shape approximation to the actual mode shape simulated by FEM is shown in Fig. 2.9. Inserting the mode shape ansatz (2.75) into Rayleigh's equation (2.7) results in the approximate eigenfrequency solution for the fundamental mode of a *clamped plate*:

$$\omega_{1,1} = 6\sqrt{2} \sqrt{\frac{7L_x^4 + 4L_x^2 L_y^2 + 7L_y^4}{L_x^4 L_y^4}} \sqrt{\frac{D_P}{\rho h}} \tag{2.77}$$

which for a square plate ($L_x = L_y = L$) simplifies to

$$\omega_{1,1} = \frac{36}{L^2} \sqrt{\frac{D_P}{\rho h}} . \tag{2.78}$$

This approximate result obtained via Rayleigh's method has an error smaller than 0.1% compared to the result obtained via FEM.

   This approximation of the fundamental eigenfrequency of a rectangular plate clamped at the boundary shows how to express the eigenfrequency more generically:

Table 2.2 Eigenfrequency constants $\alpha_{n,j}$ for a square plate fixed at the boundary [2]

(n, j)	$(1, 1)$	$(1, 2)$	$(2, 2)$	$(2, 3)$
$\alpha_{n,j}$	36.0	73.4	108.3	165.2

$$\omega_{n,j} = \alpha_{n,j} \frac{1}{L^2} \sqrt{\frac{D_P}{\rho h}}. \tag{2.79}$$

The corresponding $\alpha_{n,j}$ values can be obtained by the *Rayleigh-Ritz method*. A few values for a square plate are listed in Table 2.2.

2.3.1.2   Circular Plates

The fundamental eigenfrequency of a circular plate can readily be approximated by means of Rayleigh's method. From Table 2.1 the maximum potential and kinetic energies are given by

$$W_{pot,max} = \pi D_P \int_0^R \left\{ \left(\frac{\partial^2 U(r)}{\partial r^2} + \frac{1}{r} \frac{\partial U(r)}{\partial r} \right)^2 - 2(1-v) \frac{\partial^2 U(r)}{\partial r^2} \frac{1}{r} \frac{\partial U(r)}{\partial r} \right\} r \, dr \tag{2.80}$$

and

$$W_{kin,max} = \pi \rho h \int_0^R (U(r))^2 r \, dr. \tag{2.81}$$

The approximate mode shape function for the fundamental mode of a circular plate with clamped boundaries can be approximated by Weaver et al. [3]

$$\phi_{1,0}(r) = c \left(1 - \left(\frac{r}{R} \right)^2 \right)^2 \tag{2.82}$$

Substituting the mode shape (2.82) in (2.80) and (2.81) results in $W_{pot,max} \approx 32\pi D_p/(3R^2)$ and $W_{kin,max} \approx \pi h R^2 \rho/10$, and the eigenfrequency can be calculated with (2.7), which yields the approximation:

$$\omega_{1,0} \approx 10.33 \frac{1}{R^2} \sqrt{\frac{D_P}{h\rho}}. \tag{2.83}$$

This approximation of the fundamental eigenfrequency of a circular plate clamped at the boundary shows how to express the eigenfrequency equation more generically:

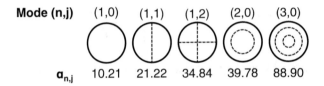

Mode (n,j) (1,0) (1,1) (1,2) (2,0) (3,0)

$a_{n,j}$ 10.21 21.22 34.84 39.78 88.90

Fig. 2.10 Mode shapes of a circular plate clamped at the boundary. n represents the circular modes and j the nodal diameters

Table 2.3 Eigenfrequency constants $\alpha_{0,j}$ for a circular plate fixed in the center for concentric modes [3]

n	$j = 0$	$j = 1$	$j = 2$	$j = 3$
1	3.75	20.91	60.68	119.7

$$\omega_{n,j} = \alpha_{n,j} \frac{1}{R^2} \sqrt{\frac{D_P}{h\rho}} \tag{2.84}$$

with the exact corresponding $\alpha_{n,j}$ values, derived via the *Rayleigh-Ritz method*, listed in Fig. 2.10.

A specific boundary condition typical for circular plates is *fixed in the center* design. Corresponding $\alpha_{n,j}$ values are given in Table 2.3.

2.3.2 Free Bending Vibration of Membranes

Plates in the tensile stress limit are typically called *membranes* or *drumheads*. The mechanical stiffness in these structures is determined by their tensile stress, in contrast to the flexural rigidity in plates as discussed before.

2.3.2.1 Rectangular Membranes

The equation of motion of a *rectangular membrane* is given by the two-dimensional wave equation [12]:

$$\sigma \nabla^2 U - \rho \frac{\partial^2 U}{\partial t^2} = 0 \tag{2.85}$$

and can be solved with the same ansatz (2.69) used for plates in Sect. 2.3.1 on page 47 of separating the variables.

With (2.70), Eq. (2.85) can be solved and the eigenfrequencies of a membrane become:

$$\omega_{n,j} = \sqrt{\frac{\sigma}{\rho}}\sqrt{\frac{n^2\pi^2}{L_x^2} + \frac{j^2\pi^2}{L_y^2}} \tag{2.86}$$

which in the case of a quadratic membranes with $L_x = L_y = L$ reduces to

$$\omega_{n,j} = \pi\sqrt{n^2 + j^2}\frac{1}{L}\sqrt{\frac{\sigma}{\rho}} = \pi\sqrt{n^2 + j^2}\frac{1}{L}c_\sigma. \tag{2.87}$$

The mode shape of a particular mode is shown in Fig. 2.11. It can be seen that the one-dimensional form of (2.87) results in the eigenfrequency of a string (2.55).

2.3.2.2 Circular Membranes

The eigenfrequency of a *circular membrane* can be readily approximated with *Rayleigh's method*, as introduced in the previous Sect. 2.3.1. The mode shape of a circular membrane is best described in circular coordinates. The fundamental mode with a vibrational amplitude c in the membrane center can then be approximated by Weaver et al. [3]

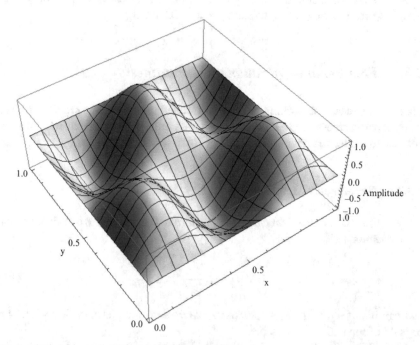

Fig. 2.11 Mode shape of a quadratic membrane for $n = 2$ and $j = 3$

$$\phi(r) = c \cos \frac{\pi r}{2R} \tag{2.88}$$

with the radial coordinate r and the membrane radius R. The maximal kinetic and potential energies are given by

$$W_{Kin,max} = \frac{1}{2}\omega^2 \rho \int_0^R \phi(r)^2 2\pi r \, dr \tag{2.89}$$

and

$$W_{Pot,max} = \frac{1}{2}\sigma \int_0^R \left(\frac{\partial \phi(r)}{\partial r}\right)^2 2\pi r \, dr, \tag{2.90}$$

respectively. An approximation for the eigenfrequency of the fundamental mode of a circular membrane can now be calculated from Rayleigh's principle (2.1), which yields

$$\omega_{1,1} \approx \frac{\pi}{2}\sqrt{\frac{\pi^2+4}{\pi^2-4}}\frac{1}{R}\sqrt{\frac{\sigma}{\rho}} = \frac{2.415}{R}\sqrt{\frac{\sigma}{\rho}}. \tag{2.91}$$

This approximation is close to the precise value of [3]

$$\frac{2.404}{R}\sqrt{\frac{\sigma}{\rho}}. \tag{2.92}$$

The term for the eigenfrequency obtained by Rayleigh's principle is qualitatively correct, and the eigenfrequencies of circular membranes can generally be expressed as

$$\omega_{n,j} = \frac{\alpha_{n,j}}{R}\sqrt{\frac{\sigma}{\rho}} = \alpha_{n,j}\frac{1}{R}c_\sigma \tag{2.93}$$

where c_σ is the velocity of a bending wave on a membrane, as introduced for strings earlier in Sect. 2.2. Values of $\alpha_{n,j}$ for the first 12 modes are listed in Fig. 2.12.

2.4 One-Dimensional Bulk Vibrations

Nanomechanical one-dimensional bulk resonators are rarer than flexural resonators, as it can be challenging to detect the small vibrational amplitudes. On the other hand, such bulk mode resonators have become the facilitating device in recent breakthrough experiments in quantum optomechanics due to their high resonance frequencies in the GHz regime.

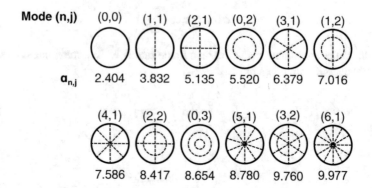

| Mode (n,j) | (0,0) | (1,1) | (2,1) | (0,2) | (3,1) | (1,2) |

$\alpha_{n,j}$ 2.404   3.832   5.135   5.520   6.379   7.016

| | (4,1) | (2,2) | (0,3) | (5,1) | (3,2) | (6,1) |

7.586   8.417   8.654   8.780   9.760   9.977

Fig. 2.12 Eigenmodes and respective $\alpha_{n,j}$ values of a circular membrane. The nodal lines (lines of zero vibration) are marked as dashed lines. n represents the straight nodal lines, and j concentric nodal circles

Fig. 2.13 Schematics of a thin beam under longitudinal vibration

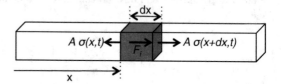

In this subsection we consider longitudinal waves in a thin beam which produces the axial displacement $U(x, t)$. Figure 2.13 shows a schematic of a thin cross-sectional piece of a thin beam. According to Newton's third law, all forces acting on the small beam piece have to equalize. The acting force is on the one hand the inertial force F_i and on the other hand the forces exerted from the neighbor piece. Assuming a thin beam with cross-sectional area A with an even axial stress distribution $\sigma(x, t)$, the force equation can be written as

$$F_i = \rho A \frac{\partial^2 U(x, t)}{\partial t^2} dx = A[\sigma(x + dx, t) - \sigma(x, t)], \qquad (2.94)$$

with the mass density ρ. For a short piece, the term in the square brackets can be approximated by

$$\sigma(x + dx, t) - \sigma(x, t) \approx \frac{\partial \sigma}{\partial x} dx \qquad (2.95)$$

and (2.94) simplifies to

$$\rho \frac{\partial^2 U(x, t)}{\partial t^2} = \frac{\partial \sigma}{\partial x}. \qquad (2.96)$$

Applying Hooke's law of elasticity, we can express the axial stress as a function of the axial deformation:

$$\sigma(x, t) = E\varepsilon(x, t) = E\frac{\partial U(x, t)}{\partial x} \tag{2.97}$$

and (2.96) turns into the one-dimensional wave equation

$$\frac{\partial^2 U(x, t)}{\partial t^2} = c_E^2 \frac{\partial^2 U(x, t)}{\partial x^2} \tag{2.98}$$

with the wave velocity inside the material of

$$c_E = \sqrt{\frac{E}{\rho}}. \tag{2.99}$$

This one-dimensional wave equation is also obtained in the case of the string resonator, as discussed in the previous Sect. 2.2.2, when dismissing the flexural rigidity in the respective equation of motion (2.47), with the appropriate wave velocity for a string (2.57).

The one-dimensional wave equation (2.98) can readily be solved with the common ansatz of separating the spatial and temporal variables, which for a individual normal mode writes

$$U(x, t) = \phi_n(x)u_n(t) = \phi_n(x)u_{0,n} \cos(\omega_n t) \tag{2.100}$$

which yields the ordinary differential equation

$$\frac{\partial^2 \phi_n}{\partial x^2} + \beta_n^2 \phi = 0 \tag{2.101}$$

with the dispersion relationship

$$\beta_n = \frac{\omega}{c_E}. \tag{2.102}$$

The resulting differential equation (2.101) can now be solved for specific boundary conditions with the general solution:

$$\phi_n(x) = a \sin(\beta_n x) + b \cos(\beta_n x). \tag{2.103}$$

In the case of both ends free condition, the stress is disappearing at the beam ends which can be expressed by

$$\frac{\partial \phi_n(0)}{\partial x} = \frac{\partial \phi_n(L)}{\partial x} = 0. \tag{2.104}$$

When applying these boundary conditions to the ordinary differential equation, (2.101) results in

$$a = 0 \quad \& \quad \sin(\beta L) = 0. \tag{2.105}$$

The latter solution is true for all wave numbers $\beta_n = n\pi/L$ with the mode numbers $n = 1, 2, 3, \ldots$. From (2.102), the eigenfrequencies can be calculated:

$$\omega_n = \frac{n\pi}{L} c_E. \tag{2.106}$$

The similarity with the eigenfrequency of a string (2.56) is now obvious. The string is fixed at both ends which results in a sinusoidal mode shape. The longitudinally vibration beam on the other hand is free at both ends, and the mode shape is given by

$$\phi_n(x) = \cos\left(\frac{n\pi}{L}x\right) \tag{2.107}$$

and depicted in Fig. 2.14.

In the case of the boundary condition that one end is free, the stress is disappearing at the beam ends which can be expressed by

$$\frac{\partial \phi_n(0)}{\partial x} = \phi_n(L) = 0 \tag{2.108}$$

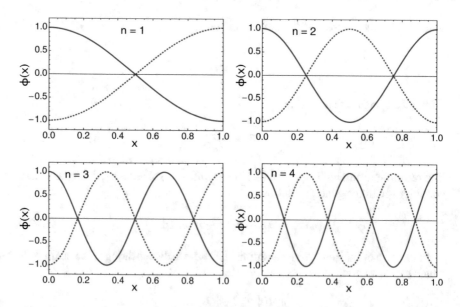

Fig. 2.14 Mode shapes $\phi_n(x)$ for a longitudinal vibration of a beam with length $L = 1$ with two free ends

the eigenfrequency becomes

$$\omega_n = \pi \frac{(2n-1)}{2} \frac{c_E}{L}.$$ (2.109)

2.5 Torsional Vibration of Thin Beams

Torsional vibrations of nanomechanical beams are a common resonator design. A few examples are shown in Fig. 2.15. The torsional vibration of a slender beam is very hard to electronically or optically transduce directly. Therefore, most torsional nanomechanical resonators feature a relatively large paddle, as schematically depicted in Fig. 2.16, which leverages the small torsional vibrations to amplitudes that can easily be actuated and detected. Typically in such torsional paddle resonators, the rotational inertia of the bars hinging the paddle can be neglected. Such resonators are best described as a lumped-element model, as discussed in Sect. 2.5.1.

The rotational eigenfrequency of a slender bar can readily be calculated by Rayleigh's method (2.7), with the potential and kinetic energies from Table 2.1:

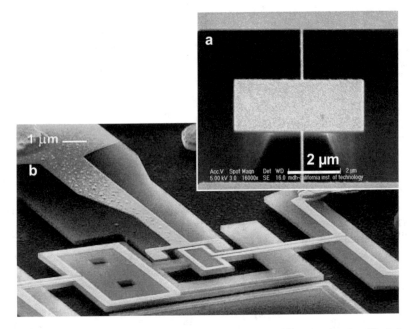

Fig. 2.15 Examples of torsional nanomechanical resonators. (**a**) Silicon torsional paddle-infrared detector. (Reprinted with permission from [13]. Copyright 2013, American Chemical Society.) (**b**) Torsional nanoelectromechanical electrometer [14]

$$W_{pot,max} = \frac{1}{2} G I_p u_0^2 \int_0^L \left(\frac{\partial \phi(x)}{\partial x} \right)^2 dx,$$ (2.110)

and

$$W_{kin,max} = \frac{1}{2} I_p \rho \omega^2 u_0^2 \int_0^L (\phi(x))^2 dx.$$ (2.111)

Assuming a sinusoidal modeshape for the case of rod clamped at both ends

$$U_n(x) = c \sin \left(\frac{n\pi}{L} x \right)$$ (2.112)

the eigenfrequency becomes:

$$\omega_n = \frac{n\pi}{L} \sqrt{\frac{G}{\rho}} = \frac{n\pi}{L} c_\varphi$$ (2.113)

with the shear wave velocity

$$c_\varphi = \sqrt{\frac{G}{\rho}}.$$ (2.114)

This eigenfrequency is apparently similar to the eigenfrequencies for strings (2.55) and one-dimensional bulk vibrations (2.106). All three structures can be fundamentally described by the one-dimensional wave equation, producing sinusoidal standing waves.

2.5.1   Torsional Paddle Resonator

The typical design of such a resonator is shown in Fig. 2.16. In this depicted case, the rotational inertia is solely concentrated in the rotating paddle in the center. The hinging bars are acting as rotational springs with a negligible rotational inertia. This assumption is particularly true for paddles hinged by carbon nanotubes.

The rotational stiffness of the connecting bars can readily be derived by comparing the potential energies of the continuum mechanical bar to the corresponding lumped-element model. Assuming a rod with constant cross section, a linear modeshape function can be used:

$$\phi(x) = \frac{x}{L}$$ (2.115)

and the potential energy of a rotational shaft (see Table 2.1)

Fig. 2.16 Schematic drawing of torsional paddle resonator

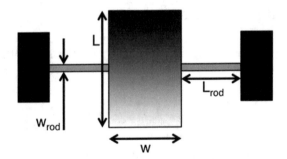

$$W_{pot,max} = \frac{1}{2}GI_p u_0^2 \int_0^{L_{rod}} \left(\frac{\partial \phi(x)}{\partial x}\right)^2 dx \qquad (2.116)$$

becomes

$$W_{pot,max} = \frac{1}{2}u_0^2 \frac{GI_p}{L_{rod}}. \qquad (2.117)$$

This energy has to be equal to the potential energy of a lumped-element sytem with $W_{pot,max} = \frac{1}{2}u_0^2 k_\varphi$, which results in the rotational stiffness of a rod of length L_{rod}:

$$k_\varphi = \frac{GI_p}{L_{rod}}. \qquad (2.118)$$

The polar geometrical moment of inertia I_p of a rod with rectangular cross section is given by

$$I_p = h^3 w_{rod} k_c. \qquad (2.119)$$

with the factor k_c, which for a square rod ($h = w_{rod}$) becomes $k_c = 0.141$, and for a thin rectangular rod ($h \ll w_{rod}$) becomes $k_c = 1/3$.

The rotational inertia of the paddle with thickness h can be calculated by Bao [15]

$$I_\varphi = \int_{-L/2}^{L/2} \rho w h x^2 dx = \frac{1}{12}\rho w h L^3. \qquad (2.120)$$

The eigenfrequency of such a rotational paddle resonator is now given by

$$\omega_0 = \sqrt{\frac{k_\varphi}{I_\varphi}}. \qquad (2.121)$$

2.6 Effective Parameters

It is commonly used to simplify the dynamics of an individual resonance mode
(normal mode) of a continuous structure with that of a harmonic resonator:

$$\omega_n = \sqrt{\frac{k_{eff,n}}{m_{eff,n}}}, \qquad (2.122)$$

where $k_{eff,n}$ and $m_{eff,n}$ denote effective spring constant and effective mass of
the particular normal mode. These effective parameters can readily be obtained
by comparing the lumped model potential or kinetic energies to the respective
continuum mechanical energies listed in Table 2.1. It is important to note that the
effective parameters depend on the chosen amplitude normalization. It is therefore
important to mention the position x chosen for the comparison. Typically, the
lateral location of the maximal displacement in the fundamental mode is chosen,
which, e.g., for a singly clamped beam is at the tip ($z_0 = u_{0,n}\phi_n(L)$), while for
a doubly clamped beam, it is in the center ($z_0 = u_{0,n}\phi_n(L/2)$). In the following
two paragraphs, two different methods to derive the effective parameters are
presented at the example of one-dimensional beams as introduced in Sect. 2.2.1. The
methodology can directly be translated for other continuum mechanical structures.
In Table 2.4 the effective mass of a few common continuum mechanical structures
is listed.

Table 2.4 Effective mass and stiffness constants of various continuum mechanical structures with
total mass m_0 for various boundary conditions BC

Member	Effective mass (m_{eff})	Normalization position x
Singly clamped beam of length L in bending vibration	$\frac{1}{4}m_0$	At tip ($x = L$)
Doubly clamped beam of length L in bending vibration	$\frac{1}{2}m_0$	Center of antinode
String of length L in bending vibration	$\frac{1}{2}m_0$	Center of antinode
Singly clamped rod of length L in longitudinal vibration	$\frac{1}{4}m_0$	At tip ($x = L$)
Doubly clamped rod of length L in longitudinal vibration	$\frac{1}{2}m_0$	Center of antinode
Rectangular plate of size S in bending vibration	$\frac{1}{4}m_0$	Center of antinode
Circular plate of radius R in bending vibration	$\frac{1}{5}m_0$	Center of antinode
Rectangular membrane of size S in bending vibration	$\frac{1}{4}m_0$	Center of antinode

2.6.1 Energy Approach

This approach is based on the comparison of the kinetic energy of the continuum mechanical structure to the kinetic energy of the corresponding lumped model system. With the kinetic energy of a one-dimensional beam resonator (see Table. 2.1), the kinetic energy comparison yields:

$$\frac{1}{2}A\rho \int_0^L \left(\frac{\partial}{\partial t}\phi_n(x)u_{0,n}\cos(\omega_n t)\right)^2 dx = \frac{1}{2}m_{eff,n}\left(\frac{\partial}{\partial t}z_0\cos(\omega_0 t)\right)^2.$$

$$(2.123)$$

Setting the absolute vibrational amplitudes of the two systems equal ($u_{0,n} = z_0$) as well as the eigenfrequencies ($\omega_n = \omega_0$), the following effective mass can be extracted:

$$m_{eff,n} = \rho A \int_0^L \phi_n^2(x)dx. \qquad (2.124)$$

Knowing the effective mass, the corresponding effective stiffness $k_{eff,n}$ can readily be obtained from the eigenfrequeny equation (1.3):

$$k_{eff,n} = \omega_n^2 m_{eff,n} \qquad (2.125)$$

or by comparing the potential energies of the equivalent systems, similar to the equalization of the kinetic energies for obtaining the effective mass.

2.6.2 Galerkin's Method

This method is a more general approach of deriving not only the effective mass but also all other effective parameters. It's based on the orthogonality of the normal modes. The derivation is performed for the case of a damped "one-dimensional" beam structure, with the differential equation of motion (2.47) extended to include an actuation force $F(x, t)$

$$\rho A\frac{\partial^2 U(x,t)}{\partial t^2} + EI_y\frac{\partial^4 U(x,t)}{\partial x^4} - \sigma A\frac{\partial^2 U(x,t)}{\partial x^2} = F(x,t). \qquad (2.126)$$

Assuming that the motion of the beam is only happening on one vibrational mode, i.e., $U(x,t) = u_n(t)\phi_n(x)$, multiplying by that mode shape $\phi_n(x)$ and integrating all over the length of the structure, i.e., $\int_0^L$, it is possible to reach the lumped-element modeling using integration by parts

$$\rho A \ddot{u}_n(t) \int_0^L \phi_n^2(x)\mathrm{d}x + E I_y u_n(t) \int_0^L \phi''^2_n(x)\mathrm{d}x + \sigma A u_n(t) \int_0^L \phi'^2_n(x)\mathrm{d}x$$

$$= \int_0^L F_n(x,t)\phi_n(x)\mathrm{d}x. \tag{2.127}$$

In doing so, all other mode components disappear:

$$\int_0^L \phi_a(x)\phi_b(x)\mathrm{d}x = 0 \quad [a \neq b], \tag{2.128}$$

except the selected mode component remains. The family of normal modes ϕ_n forms an orthogonal base for the Hilbert space of functions between 0 and L. The multiplication with a particular modeshape function and integrating over the resonator area is equivalent to taking the inner product between orthogonal norms.

From (2.127) it is possible to define the effective mass $m_{eff,n}$ (2.124), effective damping coefficient $c_{eff,n}$, effective stiffness $k_{eff,n}$, and effective force $F_{eff,n}$, so that the lumped model equation is retrieved:

$$m_{eff,n}\ddot{u}_n(t) + c_{eff,n}\dot{u}_n(t) + k_{eff,n}u_n(t) = F_{eff}(t) \tag{2.129}$$

with the effective parameters

$$m_{eff,n} = \rho A \int_0^L \phi_n^2(x)\mathrm{d}x$$

$$c_{eff,n} = c \int_0^L \phi_n^2(x)\mathrm{d}x$$

$$k_{eff,n} = E I_y \int_0^L \phi''^2_n(x)\mathrm{d}x + \sigma A \int_0^L \phi'^2_n(x)\mathrm{d}x \tag{2.130}$$

$$F_{eff,n} = \int_0^L F_n(x,t)\phi_n(x)\mathrm{d}x.$$

The presented formalism can readily be extended to different parameters (see Sect. 1.3) and to two-dimensional structures (see Sect. 6.2.1.1).

2.7 Geometric Nonlinearity

Geometric nonlinearites prominently appear in structures with fixed boundary conditions, such as doubly clamped beams and strings and their 2D equivalents such as fully clamped plates and membranes. In these structures any out-of-plane displacement produces an elongation and hence stress in-plane of the structure. It's this extra stress that causes geometric nonlinearities. The special case of doubly clamped beams is discussed in detail subsequently. The nonlinear solution for fixed plates/membranes is also given.

The here presented derivation of the nonlinear lumped-element model can be applied to any other type of nonlinearity and not only to the geometric nonlinearity. It is good to remember that the derivation has been done here for only one particular type of geometric nonlinearity, which is a nonlinearity of the effective stiffness of the mode, caused by an increase in the longitudinal tension due to motion. In other type of flexural devices, e.g., clamped-free beams or cantilevers, it can be seen that there are also inertial nonlinear terms that affect the effective mass[16–20]. Therefore, for each particular type of geometry, a slightly different analysis must be made. To conclude this paragraph, it is also important to note that the lumped model parameters are determined through the modal shapes which, as discussed at the beginning of the chapter, are in general not trivial to calculate and one might need the help of finite element modeling (FEM) to determine them.

2.7.1 Geometric Nonlinearity of Doubly Clamped Beams and Strings

The total tension in the structure of Fig. 2.17, with a Young's modulus E and a cross-sectional area A, is the sum of an existing tensile force $N_0 = \sigma A$ and a tension coming from the longitudinal strain ε of the beam during vibration (assuming small displacements):

$$N(u(x, t)) = N_0 + EA\varepsilon$$

$$\approx \sigma A + EA\frac{1}{2L} \int_0^L \left(\frac{\partial U(x, t)}{\partial x}\right)^2 dx \tag{2.131}$$

which can be then inserted in (2.47) to obtain

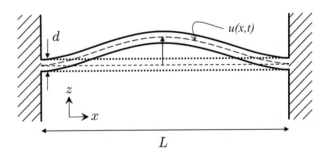

Fig. 2.17 Schematic showing the cross section of a clamped-clamped beam and how deflection induces additional strain and stress in the longitudinal axis of the beam

$$\rho A \frac{\partial^2 U(x,t)}{\partial t^2} + E I_y \frac{\partial^4 U(x,t)}{\partial x^4}$$

$$- \left[\sigma A + \frac{EA}{2L} \int_0^L \left(\frac{\partial U(x,t)}{\partial x} \right)^2 dx \right] \frac{\partial^2 U(x,t)}{\partial x^2} = 0 \qquad (2.132)$$

where the nonlinear term, the second one inside the square bracket, starts to be apparent, as it can be seen that the differential equation is not a linear function of $U(x,t)$ anymore. Applying Galerkin's method as introduced in Sect. 2.6.2 on page 61 leads to

$$\rho A \ddot{u}_n(t) \int_0^L \phi_n^2(x)dx + E I_y u_n(t) \int_0^L \phi_n''^2(x)dx + \sigma A u_n(t) \int_0^L \phi_n'^2(x)dx$$

$$+ \frac{EA}{2L} u_n^3(t) \left(\int_0^L \phi_n'^2(x)dx \right)^2 = F_n(t). \qquad (2.133)$$

From (2.133) it can be defined now the effective mass $m_{eff,n}$ (2.124), effective stiffness $k_{eff,n}$, effective damping rate $\Gamma_{eff,n}$, and nonlinear stiffness $\alpha_{eff,n}$ so that the lumped model equation is retrieved:

$$m_{eff,n} \ddot{u}_n(t) + k_{eff,n} u_n(t) + \alpha_{eff,n} u_n^3(t) = F_n(t). \qquad (2.134)$$

This equation can be also written as

$$\ddot{u}_n(t) + \frac{k_{eff,n}}{m_{eff,n}} u_n(t) + \frac{\alpha_{eff,n}}{m_{eff,n}} u_n^3(t) = \frac{F_n(t)}{m_{eff,n}}, \qquad (2.135)$$

which corresponds to the previously introduced Duffing equation (1.55) in Sect. 1.3 on page 19, with the effective Duffing parameter:

$$\alpha_{eff,n} = \frac{EA}{2L} \left(\int_0^L \phi_n'^2(x)dx \right)^2. \qquad (2.136)$$

The vibrational amplitude necessary to observe geometric nonlinearities is best discussed by rewriting the homogeneous Duffing equation (1.55) in the form:

$$\ddot{u}(t) + \omega_0^2 u(t) \left[1 + \frac{\alpha_{eff,n}}{k_{eff,n}} u^2(t) \right] = 0. \qquad (2.137)$$

Hence, geometric nonlinearites can be dismissed as $\frac{\alpha_{eff,n}}{k_{eff,n}} u^2(t) \ll 1$.

For the subsequent analysis, a sinusoidal mode shape (2.49) is used:

$$\phi(x) = \sin \left(\frac{n\pi}{L} x \right), \qquad (2.138)$$

which is an appropriate assumption for strings and simply supported beams (see Sect. 2.2.2 on page 43). This yields the effective Duffing parameter:

$$\alpha_{eff,n} = \frac{(n\pi)^4}{8} \frac{EA}{L^3} \tag{2.139}$$

and effective spring constant (2.130)

$$k_{eff,n} = \frac{EI_y(n\pi)^4}{2L^3} + \frac{\sigma A(n\pi)^2}{2L}. \tag{2.140}$$

First, let's assume a pure beam-like situation, which occurs for zero tensile stress $(\sigma = 0)$

$$\frac{\alpha_{eff,n}}{k_{eff,n}} u^2(t) = 3 \left(\frac{u(t)}{h} \right)^2. \tag{2.141}$$

This shows that for a doubly clamped beams, geometrical nonlinearities start to become significant once the vibrational amplitude $u(t)$ reaches the magnitude of the beam thickness h. In contrast, for the case of a pure string (high tension limit), the flexural rigidity term in (2.140) can be neglected, and

$$\frac{\alpha_{eff,n}}{k_{eff,n}} u^2(t) = \frac{(n\pi)^2}{4} \frac{E}{\sigma} \left(\frac{u(t)}{L} \right)^2. \tag{2.142}$$

Here, the ratio of vibrational amplitude to the lateral string length $(u(t)/L)$ is typically very small, and geometrical nonlinearities are negligible. However, it's possible that geometrical nonlinearities can be observed with strings made of stiff materials and having a low enough tensile stress. In that case the ratio E/σ can become large enough to counteract the geometrical aspect ratio.

2.7.2 Geometric Nonlinearity of Rectangular Plates or Membranes

The effective Duffing coefficient for membrane or plate resonators has been derived by [21] and is given by

$$\begin{aligned}
\alpha_{eff,n,j} &= \frac{3\pi^4}{16} \frac{E}{\rho} \left(\frac{n^4}{L_x^4} + \frac{j^4}{L_y^4} \right) m_{eff} \\
&= \frac{3\pi^4}{64} EhL_x L_y \left(\frac{n^4}{L_x^4} + \frac{j^4}{L_y^4} \right),
\end{aligned} \tag{2.143}$$

which for a square plate or membrane ($L_x = L_y = L$) reduces to

$$\alpha_{eff,n,j} = \frac{3\pi^4(n^4 + j^4)}{64}\frac{Eh}{L^2}.$$ (2.144)

2.7.3 Nonlinear Mode Coupling

The vibration of a specific mode m of, e.g., a doubly clamped beam, produces an extra tensile stress (2.131):

$$\sigma_m(t) = E\frac{1}{2L}\int_0^L \left(\frac{\partial U_m(x,t)}{\partial x}\right)^2 dx = E\frac{1}{2L}u_m^2(t)\int_0^L \left(\frac{\partial \phi_m(x)}{\partial x}\right)^2 dx.$$
(2.145)

This extra stress modulates the effective spring constant of all other modes. The effective spring constant is scaling linearly with tensile stress (2.130). Hence, the spring constant of another mode k gets modulated as follows:

$$k_{eff,k,mod}(t) = k_{eff,k}\left[1 + 2\lambda_{k,m}u_m^2(t)\right],$$ (2.146)

where $\lambda_{k,m}$ is the specific nonlinear coefficient (visit [22] for the exact derivation of $\lambda_{k,m}$ for doubly clamped beams). This effect is called *nonlinear mode coupling*. This, of course, directly affects the eigenfrequency of mode k as follows:

$$\omega_{k,mod}(t) = \omega_k\sqrt{1 + 2\lambda_{k,m}u_m^2(t)}.$$ (2.147)

Assuming that the modulation is small ($\lambda_{k,m}u_m^2(t) \ll 1$), the relative eigenfrequency change of mode k directly scales with the square of the vibrational amplitude of mode m:

$$\frac{\Delta\omega_k}{\omega_k} \approx \lambda_{k,m}u_{m,0}^2.$$ (2.148)

This quadratic behavior has been experimentally verified by means of piezoelectric nanomechanical beam resonators [22].

This analysis shows that the normal modes can couple via geometric nonlinearities. Nonlinear mode coupling results in parametric modulation of the eigenfrequency (2.147), which can produce a parametric amplification phenomenon as discussed previously in Sect. 1.4 on page 25. This effect can appear if the two modes that couple have eigenfrequencies that are whole-number fractions of each other. This can happen naturally, in particular in string and membrane resonators with harmonic higher-order modes. This effect has, e.g., been used to produce "phonon lasing" [23]. In another experiment from the same group, it has been shown that two modes can be coupled via a piezoelectrically induced stress modulation [24].

References

1. Ventsel, E., & Krauthammer, T. (2001). *Thin plates and shells: Theory, analysis, and applications*. Marcel Dekker.
2. Stokey, W. F. (2002). Vibration of systems having distributed mass and elasticity. In *Harris' shock and vibration handbook* (pp. 7.1–7.50).
3. Weaver, W., Timoshenko, S. P., & Young, D. H. (1990). *Vibration problems in engineering*. Wiley Interscience.
4. Schmid, S. (2009). *Electrostatically actuated all-polymer microbeam resonators—characterization and Application*, volume 6 of Scientific reports on micro and nanosystems. Der Andere Verlag.
5. Li, M., Myers, E. B., Tang, H. X., Aldridge, S. J., McCaig, H. C., Whiting, J. J., Simonson, R. J., Lewis, N. S., & Roukes, M. L. (2010). Nanoelectromechanical resonator arrays for ultrafast, gas-phase chromatography chemical analysis. *Nano Letters, 10*(10), 3899–3903.
6. Schmid, S., Kurek, M., & Boisen, A. (2013). Towards airborne nanoparticle mass spectrometry with nanomechanical string resonators. *SPIE Defense, Security, and Sensing, 8725*, 872525–872528.
7. Schmid, S., Jensen, K. D., Nielsen, K. H., and Boisen, A. (2011). Damping mechanisms in high-Q micro and nanomechanical string resonators. *Physical Review B, 84*(16), 165307.
8. Cleland, A. N., & Roukes, M. L. (1996). Fabrication of high frequency nanometer scale mechanical resonators from bulk Si crystals. *Applied Physics Letters, 69*, 2653.
9. Bagci, T., Simonsen, A., Schmid, S., Villanueva, L. G., Zeuthen, E., Appel, J., Taylor, J. M., Sørensen, A., Usami, K., Schliesser, A., & Polzik, E. S. (2014). Optical detection of radio waves through a nanomechanical transducer. *Nature, 507*(7490), 81–85.
10. Sayir, M. B., & Kaufmann, S. (2005). *Ingenieurmechanik 3: Dynamik*. Teubner.
11. Schmid, S., Bagci, T., Zeuthen, E., Taylor, J. M., Herring, P. K., Cassidy, M. C., Marcus, C. M., Guillermo Villanueva, L., Amato, B., Boisen, A., Cheol Shin, Y., Kong, J., Sørensen, A. S., Usami, K., & Polzik, E. S. (2014). Single-layer graphene on silicon nitride micromembrane resonators. *Journal of Applied Physics, 115*(5), 054513.
12. Timoshenko, S. P., Woinowsky-Krieger, S., and Woinowsky-Krieger (1959). *Theory of plates and shells* (2nd Ed.). McGraw-Hill.
13. Zhang, X. C., Myers, E. B., Sader, J. E., & Roukes, M. L. (2013). Nanomechanical torsional resonators for frequency-shift infrared thermal sensing. *Nano Letters, 13*(4), 1528–1534.
14. Cleland, A. N., & Roukes, M. L. (1998). A nanometre-scale mechanical electrometer. *Nature, 392*(March), 160–162.
15. Bao, M. (2005). *Analysis and design principles of MEMS devices*. Elsevier.
16. Kacem, N., Arcamone, J., Perez-Murano, F., & Hentz, S. (2010). Dynamic range enhancement of nonlinear nanomechanical resonant cantilevers for highly sensitive NEMS gas/mass sensor applications. *Journal of Micromechanics and Microengineering, 20*(4), 45023.
17. Villanueva, L. G., Karabalin, R. B., Matheny, M. H., Chi, D., Sader, J. E., & Roukes, M. L. (2013). Nonlinearity in nanomechanical cantilevers. *Physical Review B, 87*(2), 24304.
18. Dasilva, M. R. M. C. (1988). Non-linear flexural flexural torsional extensional dynamics of beams—2. Response analysis. *International Journal of Solids and Structures, 24*(12), 1235–1242.
19. Crespodasilva, M. R. M., & Glynn, C. C. (1978). Out-of-plane vibrations of a beam including nonlinear inertia and nonlinear curvature effects. *International Journal of Non-Linear Mechanics, 13*(5–6), 261–271.

20. Perisanu, S., Barois, T., Ayari, A., Poncharal, P., Choueib, M., Purcell, S. T., & Vincent, P. (2010). Beyond the linear and Duffing regimes in nanomechanics: Circularly polarized mechanical resonances of nanocantilevers. *Physical Review B, 81*(16), 165440.
21. Lu, Y., Shao, Q., Amabili, M., Yue, H., & Guo, H. (2020). Nonlinear vibration control effects of membrane structures with in-plane PVDF actuators: A parametric study. *International Journal of Non-Linear Mechanics, 122*, 103466.
22. Matheny, M. H., Villanueva, L. G., Karabalin, R. B., Sader, J. E., & Roukes, M. L. (2013). Nonlinear mode-coupling in nanomechanical systems. *Nano Letters, 13*(4), 1622–1626.
23. Mahboob, I., Nishiguchi, K., Fujiwara, A., & Yamaguchi, H. (2013). Phonon lasing in an electromechanical resonator. *Physical Review Letters, 110*(12), 127202.
24. Mahboob, I., Nishiguchi, K., Okamoto, H., & Yamaguchi, H. (2012). Phonon-cavity electromechanics. *Nature Physics, 8*(5), 387–392.

Chapter 3
Damping

The quality factor (Q) of a mechanical resonator is defined as the ratio of stored energy versus lost energy during one cycle of vibration. A high quality factor enhances the vibrational amplitude at resonance and reduces the resonance peak width. Both effects are, as will be seen in Chap. 5 on page 145, related to the precision with which the resonance frequency of a nanomechanical resonator can be detected. The smallest detectable frequency change is directly determining the sensitivity of a resonant nanomechanical sensor. A high Q is desired for applications of micro- and nanomechanical resonators, e.g., as mass sensors, frequency references, or filters, for signal processing. The total quality factor of a resonator is the sum of dissipation mechanisms:

$$\frac{1}{Q} = \frac{1}{Q_{medium}} + \frac{1}{Q_{clamping}} + \frac{1}{Q_{intrinsic}} + \frac{1}{Q_{other}}, \tag{3.1}$$

where Q_{medium} stands for all losses due to the interaction of the mechanical structure with a fluidic or ballistic medium, $Q_{clamping}$ are losses from energy radiating into the environment over the physical clamping sites of the resonator, and $Q_{instrinsic}$ summarized all dissipation mechanisms happening within the resonator, in the bulk and on the surface. Medium losses, clamping losses, and intrinsic losses are discussed in the chapter.

Q_{other} sums up all loss mechanisms that are not covered by the first three. This, e.g., includes *electrical charge damping* caused by charges trapped on the resonator. Such charges induce image charges on a nearby surface. The resulting electrostatic forces are a source of energy dissipation [1]. Another mechanism under this category is *Eddy current losses*, which results from electrical dissipation in conductive elements due to Eddy currents induced by an external magnetic field [2, 3]. This form of damping has to be considered particularly in nanomechanical resonators that are transduced with magnetic transduction techniques, as discussed in Sect. 4.1 on page 108.

S. Schmid et al., *Fundamentals of Nanomechanical Resonators*,
https://doi.org/10.1007/978-3-031-29628-4_3

3.1 Medium Interaction Losses

There are roughly three different media types by which a nanomechanical resonator can be surrounded by and which cause a energy loss. Namely, a medium can be a viscous liquid, a viscous fluidic, or a rarefied ballistic gas. In this section, the loss mechanisms of the three particular medium types are explained, and the latest damping models are listed.

3.1.1 Liquid Damping

3.1.1.1 Resonator Immersed in Liquid

The immersion of nanomechanical resonators in viscous liquids causes substantial energy loss. In the worst-case scenario, a liquid medium can overdamp the nanomechanical structure, thereby ceasing the resonance phenomenon. The effect of a viscous liquid on the quality factor of resonating cantilever beams has been theoretically modeled for bending modes [4, 5], torsional modes [6], and bending modes of beams in close proximity to a solid surface [7]. These models were developed with respect to the application in atomic force microscopy, where the vibrating microcantilever probes often are immersed in water. The models consist of numerical approximations and are not available in a short form. The behavior of a mechanical resonator in water is therefore illustrated by means of actual experimental data taken from literature; see Fig. 3.1. It can be seen that Q at the fundamental mode is of the order of unity; hence, the system is close to being overdamped. The quality factor increases steadily with increasing flexural mode. The observed trend and order of magnitude of the observed values are well predicted by the theoretical model. However, the typically very low Qs obtained in liquid are a major hindrance for applications of nanomechanical resonators directly in liquid.

Fig. 3.1 Quality factor plotted against mode number of a cantilever (500 μm long, 100 μm wide, and 1 μm thick) immersed in water. The experimental data (extracted from [8]) is compared to the theoretical model by Sader [4]

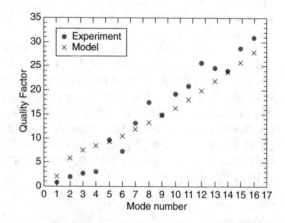

Additionally to the low quality factors, another challenge arises from the additional mass coming from the boundary liquid which is moved with the beam vibration. For example, the relative mass loading δm of a rectangular cantilever beam in the fundamental bending mode, whose length exceeds its width, immersed in a liquid is given to a good approximation by Sader [4], Chon [9], and Van Eysden and Sader [10]

$$\delta m = \frac{\pi}{4} \frac{\rho_{liquid}}{\rho} \frac{w}{h} \tag{3.2}$$

for a beam with width w, height h, and mass density ρ immersed in a liquid with mass density ρ_{liquid}. For a typical nanomechanical beam immersed in water, this results roughly in a mass load of 100%. This additional mass strongly deteriorates the mass responsivity (see Sect. 6.1 on page 178 for more details on nanomechanical mass sensing).

3.1.1.2 Liquid Inside the Resonator

The low mass responsivity together with the low quality factor results in a low sensitivity of nanomechanical resonators when immersed in a viscous liquid. A solution to this problem is to flow the liquid through the mechanical resonator instead of immersing the resonator in the liquid, as schematically depicted in Fig. 3.2. In this configuration, the resonator can be operated in vacuum, which results in almost identical quality factors for an empty or filled resonator [11]. This concept has been used for years in macroscopic U-tube density meters. More recently, micro- and nanomechanical *U-tube resonators*, also called *suspended microchannel resonators*, have been successfully used for weighing of biomolecules [11] and nanoparticles [12], and for density [13–15] and viscosity [16] measurements. The

Fig. 3.2 Schematic depiction of a U-tube resonator vibrating in vacuum and being filled with a viscous fluid

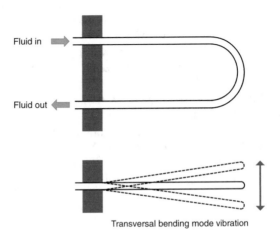

viscosity detection is nontrivial because the quality factor has been shown to be a non-monotonic function of the liquids viscosity [17, 18].

For the special case of a cantilever beam (with mass density ρ, a lot longer than thick $L \gg h$, and wider than thick $w \gg h$) comprising a fluidic channel (with mass density ρ_{liquid}, length L_{liquid}, width w_{liquid}, and height h_{liquid}), whose midplane lies on the neutral axis of the cantilever has been theoretically modeled by Sader et al. [18]. In this case, the quality factor of the fundamental bending mode resulting from the flow of an incompressible viscous liquid inside the channel is given by

$$Q_{liquid} = F(\beta_{Re}) \left(\frac{\rho}{\rho_{liquid}} \right) \left(\frac{h}{h_{liquid}} \right) \left(\frac{w}{w_{liquid}} \right) \left(\frac{L}{h_{liquid}} \right)^2 \tag{3.3}$$

with the functional

$$F(\beta_{Re}) \approx \underbrace{\frac{38.73}{\beta_{Re}}}_{\text{Small } \beta_{Re} \text{ limit}} + \underbrace{0.1521\sqrt{\beta_{Re}}}_{\text{Large } \beta_{Re} \text{ limit}} \tag{3.4}$$

which is a function of the *Reynold's number*

$$\beta_{Re} = \frac{\rho_{liquid} h_{liquid}^2}{\mu} \omega, \tag{3.5}$$

where ω is the angular frequency of the cantilever vibration and μ is the viscosity of the fluid. The functional (3.4) is the normalized quality factor; it is plotted in Fig. 3.3. It can be approximated by the sum of two extreme solutions (i) for the small $\beta_{Re} \ll 1$ limit and (ii) the large $\beta_{Re} \gg 1$ limit. From Fig. 3.3 it is obvious that the quality factor is not a monotonic function of the Reynolds number, and hence, for a fixed geometry, of the fluid viscosity.

The more intuitive behavior happens in the large Reynolds number regime, where the fluidic behavior is dominated by its inertia inside the vibrating cantilever. This inertial flow generates viscous boundary layers, which are sources of energy dissipation. As a result, an increasing fluid viscosity (decreasing β_{Re}) results in a decreasing quality factor. In contrast, the behavior in the small Reynolds number regime is a bit counterintuitive. Here, the effect of fluid inertia disappears, and the fluid behaves like a rigid body, that is, there is no generation of lossy viscous boundary layers. Hence, an increasing fluid viscosity results in an increasing quality factor. The low Reynolds number regime can be entered by reducing the fluid channel height h_{fluid}. It is thus possible to obtain highly sensitive suspended nanochannel resonators with high quality factors, which results in a high sensitivity for detecting small density changes, e.g., caused by biomolecules, cells, or nanoparticles.

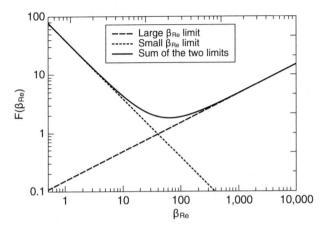

Fig. 3.3 Plot of the normalized quality factor (3.4) of a cantilever comprising an on-axis fluidic channel vs. Reynolds number β_{Re}

3.1.2 Gas Damping

Damping of a system caused by the surrounding gas is related to the surface area of the moving parts. In nanomechanical systems, the ratio of surface area to volume becomes large, and air damping can become the main source of energy dissipation. The pressure range can be divided into two regions where different damping mechanisms are dominant, namely, the fluidic and ballistic region. The transition between ballistic and fluidic regime is described by the *Knudsen number* (Kn), which is given by the ratio of the *mean free path length* of the gas (λ_f) to the representative physical length scale of the nanomechanical structure: (L_r)

$$Kn = \frac{\lambda_f}{L_r} \tag{3.6}$$

with

$$\lambda_f = \frac{k_B T}{\sqrt{2}\pi d_{gas}^2} \frac{1}{p} \tag{3.7}$$

where k_B is the Boltzmann constant, T is temperature, d_{gas} is the diameter of the gas particles, and p is the gas pressure. For air at atmospheric pressure, the mean free path is approximately 70 nm. The system is in the fluidic regime if the mean free path length of the gas is shorter than the representative length scale $\lambda_f < L_r$, that is to say $Kn < 1$. In this case, from the structure's perspective, the gas is seen as a continuum, and the interaction is best described by the fluid dynamic framework. Here, the energy dissipation results from the viscous flow of the gas around the vibrating nanomechanical structure. In contrast, the system is in the ballistic regime

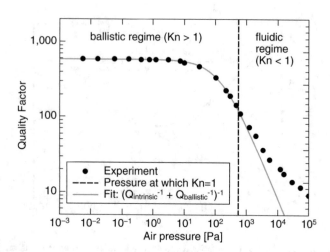

Fig. 3.4 Measured Quality factors plotted against air pressure, for a 14 μm wide SU-8 microstring. The dashed line showes the pressure $p = 550$ Pa at which $Kn = 1$, which represents the transition pressure between the fluidic and the ballistic regime. (Data extracted from [19])

if the mean free path length of the gas is larger than the representative length scale $\lambda_f > L_r$, that is to say $Kn > 1$. In this case, the gas cannot be treated as a continuous fluid. Instead, the dissipation is caused by the impact of noninteracting gas molecules.

The ballistic regime is typically reached by reducing the gas pressure. However, for small nanomechanical resonators with a representative length scale below 70 nm, the ballistic regime can even be entered at atmospheric pressure. The measured quality factor of a micromechanical string resonator as a function of air pressure is shown in Fig. 3.4. In this example, the representative length scale is the width $w = 14$ μm of the resonating string. The pressure representing $Kn = 1$ is indicated by the dashed vertical line. It can be seen that the quality factor is increasing steadily with decreasing air pressure. Until at a pressure of roughly 10 Pa, the quality factor approaches a plateau which is given by the intrinsic losses of the polymer resonator.

In the following subsections, theoretical models for the quality factor of vibrating beam structures in the two regimes are presented. For simplicity, only the two most common damping scenarios to be expected in nanomechanical resonators, namely, *squeeze-film damping* and *drag-force damping*, are covered. Nanomechanical structures fabricated by surface micromachining techniques typically are in close proximity to an underlying substrate. The gas molecules which are trapped between the substrate and the nanomechanical structure are responsible for the main gas damping, the so-called squeeze-film damping. If the vibrating beam is far away from a surface, the gas molecules are colliding with the moving surfaces of the beam which is called drag-force damping.

3.1.2.1 Fluidic Regime ($Kn < 1$)

In the fluidic regime, the dimension of the resonators is larger than the mean free path length of the gas molecules. In this case, the air can be modeled as a viscous fluid. In case the acoustic wavelength in the fluid medium (phase velocity over vibrational frequency) is larger than the representative hydrodynamic length scale of the flow (for a vibrating beam structure, this typically is the width), the models for an incompressible fluid can be applied (see Sect. 3.1.1.1 on page 70). However, for higher vibrational modes, the length scale of spatial vibration of a beam can become larger than the acoustic wavelength in the fluid. In this situation, the gas medium has to be modeled as a compressible fluid [20]. In a compressible gas, significant energy is lost via acoustic radiation, where energy is carried away in the form of sound waves.

Here, simple models based on an incompressible fluid for the squeeze-film and drag-force damping of beams are presented. These models yield simple approximations for the quality factor in a viscous gas. The quality factor due to squeeze-film (sf) damping of a long beam (width w, thickness h, mass density ρ) with a gap between the beam and the substrate of d_0 is given by Bao [21]

$$Q_{f-sf} = \frac{\rho h d_0^3}{\mu w^2} \omega \tag{3.8}$$

where μ is the coefficient of viscosity of the fluid at a given temperature.

The drag-force damping can be approximated by a dish-string model. The drag force of a disk can be analytically modelled. The simplified dish-string model replaces the beam with a string of dishes with the diameter equal to the beam width. The quality factor due to viscous drag forces (df) is then given by Bao [21]

$$Q_{f-df} = \frac{\rho w h}{8 \mu} \omega. \tag{3.9}$$

As a main rule, the quality factor due to viscous damping increases linearly with the vibrational frequency. This effect can be observed, e.g., in the measured quality factors of silicon nitride string resonators with different resonance frequencies as shown in Fig. 3.5. When comparing (3.8) and (3.9)

$$\frac{Q_{f-sf}}{Q_{f-df}} = 8 \left(\frac{d_0}{w} \right)^3 \tag{3.10}$$

it becomes obvious that squeeze-film damping drastically exceeds the effect of drag-force damping if the gap distance is smaller than the beam width ($d_0 < w$).

For both nano- [22] and microresonators [23], the influence of geometry has shown to be more complex than depicted by these simplified models for viscous damping. It seems that there is an optimal beam width at which air damping is minimal (see Fig. 3.5). Such effects come from the compressibility of the gas.

Fig. 3.5 Quality factor of silicon nitride strings as a function of the string width w measured at atmospheric pressure [23]

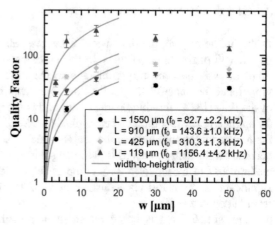

Fig. 3.6 Schematic drawing of a moving plate in rarefied gas before and after the collision with a gas molecule

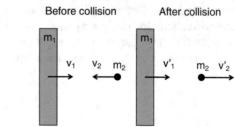

3.1.2.2 Ballistic Regime ($Kn > 1$)

Gas damping in the ballistic regime is based on momentum transfer from the resonator to colliding gas molecules. The energy transfer models are based on rigid oscillating plates, as schematically depicted in Fig. 3.6. The conservation of momentum and kinetic energy during a collision for a forward movement of the plate leads to a system of equations:

$$m_1 v_1 - m_2 v_2 = m_1 v'_1 + m_2 v'_2$$
$$\frac{1}{2} m_1 v_1^2 + \frac{1}{2} m_2 v_2^2 = \frac{1}{2} m_1 v'^2_1 + \frac{1}{2} m_2 v'^2_2. \quad (3.11)$$

Solving for the velocity of the gas molecule after the collision under the assumption that $m_1 \gg m_2$ gives

$$v'_2 = v_2 + 2 v_1. \quad (3.12)$$

So, the gas molecule is faster and hence has gained energy after the collision. However, if the plate would move backward, in direction of the molecule, the velocity of the gas molecule after the collision would be

$$v'_2 = v_2 - 2v_1, \tag{3.13}$$

and hence the molecule would have transferred energy to the plate. At first sight it seems that the total energy of the plate remains constant during vibration, as the energy transfer with the colliding gas molecules seems balanced. However, the net energy loss comes from the fact that the number of collisions at the front side is larger than the number of collisions on the back side. In other words, the "pressure" on the front side is larger than on the back side. Based on a statistical analysis of the number collisions with gas molecules, the following formula has been derived for the drag-force damping of an oscillating plate in the ballistic regime [24, 25]:

$$Q_{b-df}(p) = \frac{\rho h \omega}{4}\sqrt{\frac{\pi}{2}}\sqrt{\frac{R_{gas}T}{M_m}}\frac{1}{p} \tag{3.14}$$

where ω, T, M_m, and R_{gas} are the angular velocity of the oscillation, temperature, molar mass of the gas, and the universal molar gas constant, respectively. This air damping model for an isolated plate has been shown to also be valid for flexible beams [26]. Dry air has an average molar mass of 28.97 g/mol.

A model for the squeeze-film damping in the ballistic regime has been obtained based on similar considerations. For a beam with a peripheral length L_p, thickness h, and a distance d_0 to the substrate and a mass density ρ, the quality factor for squeeze-film damping in the ballistic regime becomes [24]

$$Q_{b-sf}(p) = (2\pi)^{\frac{3}{2}}\rho h \omega \frac{d_0}{L_p}\sqrt{\frac{R_{gas}T}{M_m}}\frac{1}{p}. \tag{3.15}$$

Both the drag-force damping and the squeeze-film damping in the ballistic regime are linear functions of the gas pressure. This linear behavior is visible in Fig. 3.4, until the measured quality factor reaches the intrinsic plateau. The quality factor as a function of gas pressure can thus be represented by the following function, which is plotted as a gray line in Fig. 3.4:

$$Q^{-1} = Q^{-1}_{intrinsic} + c_1 p \tag{3.16}$$

with the fit parameter c_1. It further can be observed that the experimental data starts to deviate from the fit exactly at the critical pressure (dashed line) when entering the fluidic regime, where the ballistic model ceases to be correct.

3.2 Clamping Loss

Clamping losses are caused by radiation of vibrational energy through the anchor of a micro- or nanomechanical resonator. The energy loss over the clamping

Fig. 3.7 Schematic
representation of the
clamping area of a nanobeam
resonator [27]

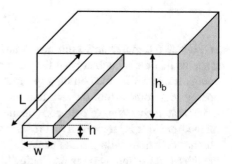

is strongly dependent on the detailed anchor geometry. Even though there are
multiple analytical models taking account for different anchor geometries, practically, the anchor geometry of a real nanomechanical oscillator differs slightly
of the theoretical scenarios. Nevertheless, the qualitative understanding based on
analytical clamping loss models helps in designing of nanomechanical resonators
with minimal clamping losses. Here, specific clamping loss models for cantilever
beams and membrane resonators are briefly introduced.

3.2.1 Cantilever Beams

Clamping loss of singly clamped cantilever beams of the geometry shown in
Fig. 3.7 has been analytically modeled by Photiadis et al. [27]. In the case that the
supporting substrate thickness is thinner than the propagating elastic waves, and
further assuming that the material properties of the base and cantilever are the same
and taking $\nu = 0.3$, clamping losses for the fundamental mode can be estimated by
the following expression:

$$Q_{clamping}^{-1} \approx 0.95 \frac{w}{L} \frac{h^2}{h_b^2} \tag{3.17}$$

where w is the width and h_b is the thickness of the supporting substrate. The $Q \propto$
L/w proportionality has also been derived for doubly clamped beams ending into a
wide plate of the same thickness ($h_s = h$) by Cross and Lifshitz [28].

   In the case that the supporting structure is sufficiently thick relative to the
wavelength of propagating waves, the base may be approximated as a semi-infinite
elastic medium. Again, assuming that the material properties of the base and
cantilever are the same and taking $\nu = 0.3$, the clamping loss expression changes to

$$Q_{clamping}^{-1} \approx 0.31 \frac{w}{L} \left(\frac{h}{L}\right)^4 . \tag{3.18}$$

As can be seen from (3.17) and (3.18), slight changes in the boundary conditions result in big changes in the geometrical dependence of clamping loss. From these two specific cases, it is possible to derive a few general rules in order to minimize clamping losses, also for doubly clamped beams. The comparison of (3.17) and (3.18) makes it clear that clamping losses are significantly reduced with a thick body chip which mechanically behaves like an semi-infinite substrate. But in both cases (plate substrate and semi-infinite substrate), clamping losses are reduced for long and slender beams. In this case the mechanical coupling into the substrate of the vibrational energy is minimized.

Clamping losses can also be determined by means of the finite element method (FEM). The computed energy that is lost over the anchors is compared to the energy stored in the system. In order to lose energy in a FEM simulation, it is assumed that all elastic waves radiating into the substrate would not be reflected and thus would be lost. Therefore, the substrate needs to be modeled as a semi-infinite domain which requires the so-called perfectly matched layer (PML) at the boundaries. A proper PML absorbs all incoming waves, producing a quasi-infinite extension of the substrate. Figure 3.8 shows the simulated displacement field of a microcantilever with an anchor terminated with a PML which absorbs all incoming elastic waves.

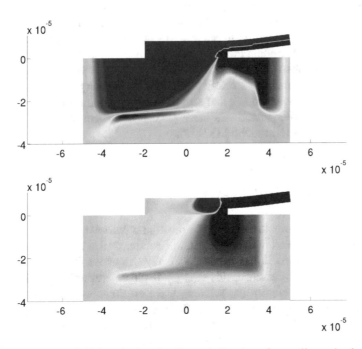

Fig. 3.8 Displacement field in x (top) and y (bottom) direction of a cantilever simulated with HiQLab. The anchor is modeled as a semi-infinite domain by terminating it with a perfectly matched layer (PML). Red areas experience a positive displacement and blue areas a negative displacement [29]

3.2.2 Membranes

Another resonator design for which a clamping loss model has been derived are membranes. An analytical model based on the coupling of membrane modes to free modes of a semi-infinite substrate has been fully developed [30, 31]. The asymptotic limit for a square membrane is given by

$$Q_{clamping} \approx 1.5 \frac{\rho_s}{\rho_r} \eta^3 \frac{n^2 m^2}{(n^2 + m^2)^{3/2}} \frac{L}{h} \tag{3.19}$$

with the "acoustic mismatch" (phase velocity ratio) between a semi-infinite substrate and the resonator

$$\eta \approx \sqrt{\frac{E_s}{\rho_s} \frac{\rho_r}{\sigma}} \tag{3.20}$$

with the mass densities ρ_s and ρ_r of the substrate and resonator, respectively, and the Young's modulus of the substrate E_s. Equation (3.19) is valid under the condition $n, m \gg \sqrt{n^2 + m^2}/\eta$. Typically, $\eta \gg 1$ for, e.g., silicon nitride membranes and thus the radiation loss model is valid for all $n \sim m$. Destructive interference of the waves radiating into the substrate can lead to a suppression of the acoustic radiation loss for increasing harmonic modes ($n = m$) [30]. From (3.19) it can be seen that acoustic radiation loss is minimal for harmonic modes $n = m$, and the envelope of maximal values is increasing linearly with the mode numbers $Q_{clamping} \propto n$.

It has repeatedly been shown that especially lower mode Qs are sensitive to the chip mounting conditions and that these Qs can be increased by minimizing the contact between chip and support [32–35].

3.2.3 Strategies Against Clamping Losses

Following the previous discussion, the most straightforward approach to reduce clamping losses in bending mode vibrations is to maximize the length to thickness aspect ratio (L/h), as it has been specifically discussed for beams (3.18) and membranes (3.19). In essence, this strategy is closely related to maximizing the acoustic mismatch between the standing wave on the nanomechanical resonator with respect to the supporting substrate. For example, in the case of nanomechanical drumheads, according to (3.20), this can readily be achieved by lowering the tensile stress of the resonator, which has shown to be an effective strategy in silicon nitride drumheads [36]. Besides bending mode vibrations, improving and controlling the acoustic mismatch is also of importance for bulk modes, e.g., for resonator piezoelectric contour mode resonators [37].

Another strategy is to minimize the coupling of the nanomechanical resonator mode vibrations to the substrate by clamping the resonator at its nodal points. At these points the vibrational amplitudes are close to zero, and hence only minimal elastic energy can propagate through the anchors. This has been successfully demonstrated for bulk mode [38] as well as bending mode resonators [39].

Finally, a very successful way to efficiently suppress radiation losses is to locate the mechanical structure within a well-designed *phononic crystal* structure. The periodicity of the elastic properties is typically created by iterating areas of resonator material and vacuum/air. This is either done by subtracting material, e.g., by punching periodic hole patterns, or by adding material, e.g., by corrugating a nanomechanical beam. Such periodic structures exhibit a frequency range in which elastic waves can't propagate, hence creating a *phononic band gap*. One strategy is to anchor a nanomechanical resonator element with such two-dimensional periodic structures. The band gap frequency of the phononic crystal has to be designed to match the specific resonator mode frequency. Then, the probability of phonons leaking off the resonator into the anchor structures is strongly suppressed. Such phononic crystal clampings have been successfully implemented to support SiN drumheads [40, 41] or microbar bulk mode resonators [42].

The most effective approach however is to localize waves by introducing a defect directly inside such a two-dimensional periodic structure. If the resonant mode frequency of such a defect area is matched with the band gap frequency of the surrounding phononic crystal, no elastic waves can propagate away from the defect, resulting in high-Q and high-frequency nanomechanical resonators. Such phononic band gap schemes have, e.g., been successfully used to localize bulk modes inside a two-dimensional piezoelectric plate [43, 44] or a one-dimensional nanomechanical silicon slab [45, 46]. In the latter example, the phononic structures were designed such that the defect can co-localize phononic together with photonic modes [47], resulting in a strong optomechanical coupling. Similar phononic crystal designs, without the photonic co-localization, have been realized in nanomechanical drumhead resonators [48, 49], creating a so-called *soft clamping*, which will be discussed under the framework of dissipation dilution in Sect. 3.3.3 on page 92.

3.3 Intrinsic Loss

In this section, all energy loss mechanisms that take place within, that is, on the surface or in the bulk of the material of the resonating structure, are summarized. The loss mechanisms can be divided into two categories: *friction loss* and *fundamental loss*. The first category is based on the friction coming from material imperfections that happen in the bulk and on the surface of a resonator, and it is discussed subsequently in Sect. 3.3.1 on the next page. The latter category are damping mechanisms constituting the fundamental loss limits present even in an ideal frictionless material. These fundamental loss mechanisms are based on interactions between the strain field inside the resonator during vibration and phonons and electrons and are discussed subsequently in Sect. 3.3.2 on page 90. In the last Sect. 3.3.3 on page 92, *dissipation dilution* is discussed, an effect arising in mechanical resonators under tensile stress that results in enhanced quality factors.

3.3.1 Friction Losses

Friction losses origin from irreversible motion of atoms during vibration. This atomic motion can be caused by, e.g., defect dislocations in crystalline materials, grain boundary slipping in metals, phase boundary slipping in bilayer structures, or molecular chain movement in amorphous materials. Friction loss mechanisms are a ubiquitous phenomenon and are best described by the anelasticity framework.

Materials which exhibit both elastic and viscous, due to friction losses, behaviors are called *viscoelastic*. A special type of viscoelastic materials that do not show any lasting deformation is called *anelastic*. There are various models to describe the anelastic response. These models are often represented by "spring-dashpot" combinations which help model materials with multiple friction loss mechanisms. The *standard linear solid* (SLS) model (see Fig. 3.9a), also called *Zener model*, is often used to describe a material with a single frequency- dependent friction loss mechanism (or relaxation mechanisms). Figure 3.9b shows the response of an anelastic material, represented by the SLS model, to a constant stress σ_0. From the response it can be seen that the induced strain ε is lagging behind the applied stress. From an immediate unrelaxed strain response ε_u, the material creeps exponentially with the relaxation time τ_σ until it reaches the relaxed strain ε_r.

From Fig. 3.9b it is obvious that an oscillatory displacement-induced stress $\sigma(t)$ and the accompanying oscillating strain $\varepsilon(t)$ are not perfectly in phase. If the strain lags behind the stress by a phase δ for an oscillatory stress, we write:

$$\varepsilon(\omega) = \varepsilon_0 \sin(\omega t)$$
$$\sigma(\omega) = \sigma_0 \sin(\omega t + \delta). \tag{3.21}$$

Using the appropriate trigonometric identity, the stress can be expanded to

$$\sigma(\omega) = \sigma_0 \sin \omega t \, \cos \delta + \sigma_0 \cos \omega t \, \sin \delta. \tag{3.22}$$

It can be seen that the stress consists of two components: one of magnitude $(\sigma_0 \cos \delta)$, in phase with the strain, and the other of magnitude $(\sigma_0 \sin \delta)$, $\pi/2$ out of phase with the strain:

The stress-strain relationship can therefore be defined by a quantity E' in phase with the strain and by a quantity E'', $\pi/2$ out of phase with the strain.

$$\sigma = \varepsilon_0 E' \sin \omega t + \varepsilon_0 E'' \cos \omega t \tag{3.23}$$

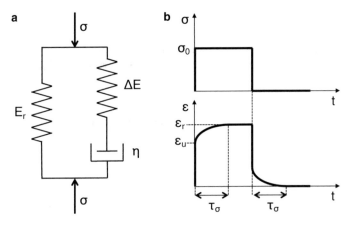

Fig. 3.9 (**a**) Spring-dashpot model of standard linear solid (SLS), also known as Zener model. (**b**) Time response of SLS to constant stress

where

$$E' = \frac{\sigma_0}{\varepsilon_0} \cos\delta \ , \quad E'' = \frac{\sigma_0}{\varepsilon_0} \sin\delta \qquad (3.24)$$

E' and E'' are the real and an imaginary part of the complex Young's modulus $E^*(\omega) = E'(\omega) + iE''(\omega)$, respectively. E' is called *storage Young's modulus* and defines the energy stored in the specimen due to the applied strain. E'' is the *loss Young's modulus* and represents the dissipation of energy.

Since the damping of a linear resonator typically is measured via the quality factor Q (see Sect. 1.1), it is of interest to express the relationship between the intrinsic material damping to Q. According to the definition (1.26), Q is the ratio of the total stored energy over the energy lost during one cycle of oscillation.

The energy (ΔW) dissipated per cycle in a volume unit of material can be calculated by

$$\Delta W = \oint \sigma \, d\varepsilon = \int_0^{2\pi/\omega} \sigma \frac{d\varepsilon}{dt} dt. \qquad (3.25)$$

Substituting for σ and ε, this term becomes:

$$\Delta W = \omega \varepsilon_0^2 \int_0^{2\pi/\omega} (E' \sin(\omega t) \cos(\omega t) + E'' \cos^2(\omega t)) dt. \qquad (3.26)$$

The integral is solved by using the trigonometric identities $\sin \omega t \cos \omega t = \frac{1}{2} \sin 2\omega t$ and $\cos^2 \omega t = \frac{1}{2}(1 + \cos 2\omega t)$, which results in

$$\Delta W = \pi E'' \varepsilon_0^2. \tag{3.27}$$

The maximum energy stored in a unit volume of material can be calculated by integrating the first term in (3.26) only over a quarter cycle rather than over the complete period:

$$W = E' \omega \varepsilon_0^2 \int_0^{\pi/(2\omega)} \sin(\omega t) \cos(\omega t) dt \tag{3.28}$$

which evaluated as before gives

$$W = \frac{1}{2} E' \varepsilon_0^2. \tag{3.29}$$

Substituting (3.27) and (3.29) in the definition of Q (1.26) yields

$$Q_{friction} = 2\pi \frac{W}{\Delta W} = \frac{E'}{E''}. \tag{3.30}$$

This equation can further be extended by the definitions for E' and E'' (3.24) which gives the definition of $\tan \delta$ also called the *loss tangent*:

$$Q_{friction} = \frac{E'}{E''} = (\tan \delta)^{-1}. \tag{3.31}$$

Solid materials typically exhibit relatively low intrinsic damping $Q > 100$ or $\tan \delta < 0.01 \approx \delta$. Hence, the dynamic mechanical characteristic can be defined in terms of the Young's modulus $\|E^*\| \approx E' = E$ and the quality factor due to intrinsic material damping Q_{intr}. Figure 3.10 shows the loss tangent as a function of the Young's modulus for a variety of different materials. It can clearly be seen that the materials typically used in micro- and nanofabrication show a combination of high Young's modulus and low damping.

Anelastic materials can exhibit a strongly frequency-dependent behavior. Furthermore, materials can have multiple relaxation mechanisms at different frequencies. The dynamic behavior of the standard linear solid, representing a single relaxation mechanism, can be calculated and the dynamic modulus is given by Fantozzi [51]

$$E'(\omega) = E_r + \Delta E \frac{\omega^2 \tau_\varepsilon^2}{1 + \omega^2 \tau_\varepsilon^2}$$
$$E''(\omega) = \Delta E \frac{\omega \tau_\varepsilon}{1 + \omega^2 \tau_\varepsilon^2} \tag{3.32}$$

with

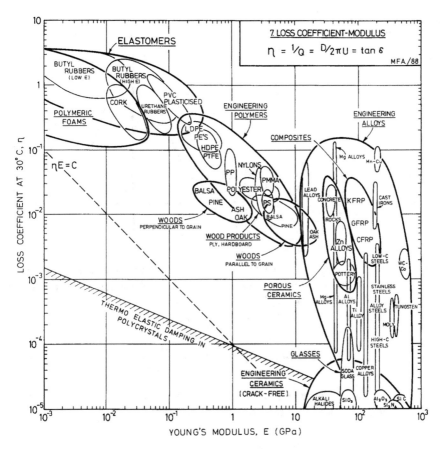

Fig. 3.10 Loss factor ($\eta' \approx \tan\delta \approx Q^{-1}_{friction}$) of various common materials (Reprinted from [50], with permission from Elsevier)

$$\tau_\varepsilon = \frac{\eta}{\Delta E} \tag{3.33}$$

being the relaxation time at constant strain with the damping constant η (see dashpot in Fig. 3.9a). In contrast, the relaxation time at constant stress, as shown in Fig. 3.9b, is given by

$$\tau_\sigma = \tau_\varepsilon \frac{E_r + \Delta E}{E_r}. \tag{3.34}$$

These equations are called the *Debye equations* (they were first derived by Debye for the dielectric relaxation phenomena). The qualitative behavior of E' and E'' as functions of $\omega\tau_\varepsilon$ is shown in Fig. 3.11. E' is undergoing the transition from the *unrelaxed Young's modulus* $E_u = E_r + \Delta E$ at ($\omega\tau_\varepsilon \gg 1$) to the *relaxed Young's*

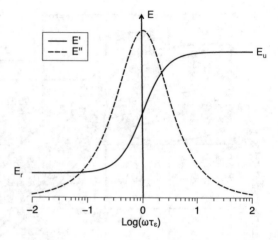

Fig. 3.11 Variation of dynamic modulus versus $\omega\tau_\varepsilon$

modulus E_r at ($\omega\tau_\varepsilon \ll 1$). Hence, the term relaxation mechanism is often used here equivalently to loss mechanism. The curve E'' has a maximum for $\omega\tau_\varepsilon = 1$ at which $E''_{max} = \frac{1}{2}\Delta E$. The shape of E'' is referred as a *Debye peak*.

The internal friction $\tan\delta$ can be obtained using (3.31)

$$\tan\delta(\omega) = \frac{E''}{E'} = \Delta E \frac{\omega\tau_\varepsilon}{E_r + E_u\omega^2\tau_\varepsilon^2}. \tag{3.35}$$

The internal friction varies as a Debye peak. Since E_r and E_u are of the same order of magnitude and a lot larger than ΔE, (3.35) can be written as

$$Q^{-1}_{friction} = \tan\delta(\omega) = \Delta \frac{\omega\overline{\tau}}{1 + \omega^2\overline{\tau}^2} \tag{3.36}$$

with the so-called relaxation strength

$$\Delta = \frac{\Delta E}{\sqrt{E_r E_u}}, \tag{3.37}$$

where τ_ε was replaced by the geometrical average of the strain at constant strain and stress $\overline{\tau} = \sqrt{\tau_\sigma\tau_\varepsilon}$ for generalization purposes. The internal friction has a maximum at $\omega\overline{\tau} = 1$, and the peak height is related to the relaxation strength Δ:

$$Q^{-1}_{min} = [\tan\delta(\omega)]_{max} = \frac{1}{2}\Delta \tag{3.38}$$

Dissipation peaks of this form (Debye peaks) are quite ubiquitous. They occur for many different relaxation mechanisms but also for phonon-phonon interaction and thermoelastic relaxation, as discussed in the next subsection.

In many of these examples, there is not just one single relaxation time $\bar{\tau}$, and therefore one sees multiple or broadened Debye peaks. One can understand qualitatively why there is a peak in dissipation when $\omega\bar{\tau} = 1$ in the following way: If the ω is much smaller than the effective relaxation rate ($\omega \ll 1/\bar{\tau}$) of the solid, then the system remains essentially in equilibrium and very little energy is dissipated. This regime is comparable to an isothermal process in thermodynamics. If the vibration frequency is much larger than the effective relaxation rate ($\omega \gg 1/\bar{\tau}$), the system has no time to relax and again very little energy is dissipated. This regime corresponds to an "adiabatic" process. It is only when the vibration frequency is on the order of the system's effective relaxation rate that appreciable dissipation occurs (3.38). The full picture may be more complicated.

3.3.1.1 Time-Temperature Equivalence

Besides being frequency-dependent, as shown in this section, relaxation mechanisms are also dependent on temperature. The principle of time-temperature equivalence implies that a specific relaxation mechanism in the material can equally be crossed by changing either the frequency or temperature. The damping behavior at one temperature can be related to that at another temperature by changing the time scale, that is, the frequency, as illustrated in Fig. 3.12. The two temperature responses are connected via the *shift factor* a_T.

During a material relaxation, the storage modulus, as depicted in Fig. 3.11, is crossing between two energy levels E_r and E_u. Hence, such a transition can be modeled as a *two-level system*, in analogy to the two energy levels, e.g., of a chemical reaction. The jump between the levels requires the activation energy ΔH. Such a transition with a constant activation energy can then be described by *Arrhenius' law*[52]:

$$\omega = \omega' \, e^{-\frac{\Delta H}{k_B T}}. \tag{3.39}$$

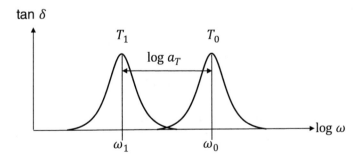

Fig. 3.12 Schematic illustration of the time-temperature equivalence

With (3.39), the shift factor a_T, introduced in Fig. 3.12, can now readily be calculated:

$$\log a_T = \frac{\log \omega_0}{\log \omega_1}$$

$$= \frac{\Delta H}{k_B} \left(\frac{1}{T_1} - \frac{1}{T_0} \right). \tag{3.40}$$

With the shift factor, it is possible to describe the Debye peak (3.36) as a function of temperature:

$$Q^{-1}_{friction}(T_1) = \tan \delta = \frac{\Delta E}{E} \frac{\omega_0 \tau_\varepsilon \exp \left\{ \frac{\Delta H}{k_B} \left(\frac{1}{T_1} - \frac{1}{T_0} \right) \right\}}{1 + \omega_0^2 \tau_\varepsilon^2 \exp \left\{ 2\frac{\Delta H}{k_B} \left(\frac{1}{T_1} - \frac{1}{T_0} \right) \right\}} \tag{3.41}$$

A transition through a Debye peak in the frequency and the temperature domain is visualized in Fig. 3.13. This shows that a material relaxation mechanism can be crossed either by varying the frequency or the temperature. In the example the frequency was fixed below the peak maximum. The transition can then be crossed by lowering the temperature. For relatively large activation energies, small changes in temperature (within one order of magnitude) have the same effect as large changes of frequency (over four orders of magnitude). Hence, to measure material transitions, it is a common technique to fix the oscillation frequency and ramp the temperature.

From Fig. 3.13 it becomes further evident that specific damping mechanisms can be frozen out. With decreasing temperature, after the Debye peak transition, the loss tangent diminishes quickly.

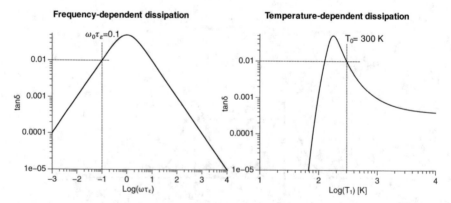

Fig. 3.13 Schematic illustration of a Debye peak in the frequency and the temperature domain, for a reference temperature $T_0 = 300K$ at $\omega_0 \tau_\varepsilon = 0.1$, damping strength $\Delta E/E = 0.1$, and activation energy $\Delta H = 1000 k_B$

3.3.1.2 Surface Friction

Surface friction loss can be the dominant source of damping in submicrometer thick beams when the surface-to-volume ratio increases. Surface loss is assumed to be caused by adsorbates on the surface [53], surface roughness [54], surface oxidation [55], or surface impurities, e.g., as a result from fabrication processing [36]. Surface loss has, e.g., been found to be the dominating damping mechanism in nanomechanical silicon nitride cantilevers [56], silicon cantilevers [55], and strings and membranes [36]. An estimate of surface loss is given by Yang et al. [53], and Yasumura et al. [56]

$$Q_{surface} = \frac{wh}{2\delta_S(3w+h)}\frac{E}{E_S''} \tag{3.42}$$

where δ_S and E_S'' are the thickness and the loss Young's modulus of the lossy surface layer and w is the beam width. For a wide beam ($w \gg h$), surface damping becomes proportional to the inverse of the beam thickness:

$$Q_{surface} = \frac{1}{6\delta_S}\frac{E}{E_S''} \cdot h. \tag{3.43}$$

This linear proportionality with the resonator thickness has, e.g., been observed with silicon nitride membranes of various thickness, as shown in Fig. 3.14.

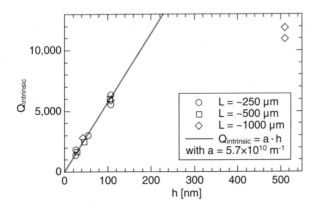

Fig. 3.14 Intrinsic quality factors of silicon-rich silicon nitride membranes as a function of membrane thickness. Data taken from [36]

3.3.2 *Fundamental Losses*

In this subsection, two particular fundamental dissipation mechanisms are introduced, namely, thermoelastic loss and phonon-phonon interaction loss (also called Akhiezer damping). Both mechanisms are coming from nonreversible heat flow inside a solid under period deformation. The heat flow (thermal relaxation) generated entropy which causes loss of energy. In thermoelastic damping the thermal relaxation happens between strain-induced temperature differences in different spatial areas in the solid. In contrast, in Akhiezer damping the thermal relaxation happens between strain-induced temperature difference in normal modes of the atomic lattice.

Similar to the description of friction losses discussed in the previous subsection, the two particular fundamental losses discussed in this subsection can be modeled with Zener's approach as Debye peaks [see (3.36)]:

$$Q^{-1}(\omega) = \Delta \frac{\omega\tau}{1 + (\omega\tau)^2} \tag{3.44}$$

with the particular relaxation strength Δ and relaxation time τ. The maximal loss of $\frac{1}{2}\Delta$ is reached when $\omega\tau = 1$. In contrast to friction losses, the relaxation strength and relaxation time for the fundamental losses can be modeled precisely. Also, where the temperature dependence of friction loss mechanisms is described by a chemical approach, by means of Arrhenius' law, fundamental losses have a known linear temperature dependence. Both mechanisms discussed subsequently linearly decrease with temperature.

Additionally to the two loss mechanisms discussed in the subsection, there are other fundamental mechanisms. For example, in materials with free electrons, such as metals and semiconductors, elastic waves can also interact with the electron cloud. Acoustic waves can induce forces on electrons. Energy is dissipated by Joule heating from the resistive electron movement. This so-called phonon-electron interaction loss in conductor will not be discussed here. More details can be found elsewhere [57].

3.3.2.1 Thermoelastic Damping

Strain due to the mechanical motion generates differences in the temperature at different locations in the resonator. The strain field is coupled to the temperature field by the material's thermal-expansion coefficient α_{th} of the resonator material. The locally generated heat flow between points of different temperatures is irreversible, that is, it creates entropy, which results in energy dissipation. This intrinsic damping mechanism is called thermoelastic damping, and was measured in micromechanical silicon resonators for the first time by Roszhart in 1990 [58].

During transverse flexural vibration of an unstressed beam, one side is under compression while the other side is under tension. The side under compression becomes warmer while the side with tension becomes colder. This temperature gradient across the beam thickness causes the energy loss. The thermoelastic loss for a beam under transverse flexural vibration has been solved by Zener in the 1940s [59] and is approximately given by

$$\Delta_{TED} = \frac{E\alpha_{th}^2 T_0}{\rho c_p} \tag{3.45}$$

and the relaxation time

$$\tau_{TED} = \frac{h^2}{\pi^2 \chi} \tag{3.46}$$

where $\chi = \kappa/(\rho c_p)$ is the thermal diffusivity of the material. κ and c_p are the thermal conductivity and the specific heat capacity, respectively. τ_{TED} is the characteristic time required for the heat to diffuse across the beam thickness. The maximum energy in an oscillation cycle is lost if the cycle time of the resonating beam is similar to τ_{TED} which is the time the heat needs to relax over the beam thickness. For lower vibration frequencies, the system stays more or less in equilibrium (isothermal state), and only little energy is lost. If the frequency of vibration is higher than the relaxation rate of the heat flow, the heat has no time to relax (adiabatic state) and again very little energy is lost.

The exact expression for thermoelastic damping in vibrating thin beams was obtained by Lifshitz and Roukes in 2000 [60], which has shown that Zener's model is an accurate approximation. Thermoelastic damping has also been modeled for bilayered beams [61].

3.3.2.2 Phonon-Phonon Interaction Loss (Akhiezer Damping)

Phonon-phonon interaction loss is a fundamental source of dissipation in crystalline material. The loss comes from the *Akhiezer effect*, which is the interaction of an oscillating low-frequency strain field with the existing high-frequency atomic thermal motion in a crystal lattice. The oscillating deformation periodically dilates the atomic crystal. The dilatation changes the elastic constant in the crystal, which changes the normal mode frequencies of the atoms. The effect of the dilatation on the normal mode frequencies is described by the Grüneisen constant γ_G. A positive dilatation (stretching) causes a frequency decrease (reduction of the temperature), and a negative dilatation (compression) causes a frequency increase (increase in temperature). However, the oscillating strain does not act on all normal modes equally. Hence, a deformation results in a temperature difference between different normal modes. During a periodic low-frequency deformation, the modes with different temperatures relax toward the mean temperature value. In this way, energy is lost by an increase of entropy from the heat flow between phonon modes with different temperatures.

Under the assumption that the strain field frequency ω is lower than the thermal phonon frequency (higher THz regime), Akhiezer damping can be modeled by Zener's approach with a Debye peak with the relaxation strength [57]:

$$\Delta_{Akhiezer} = \frac{c_p \rho T_0 \gamma^2}{E} \tag{3.47}$$

and the thermal relaxation time

$$\tau_{Akhiezer} = \frac{3\kappa}{c_p E}. \tag{3.48}$$

The intra-mode relaxation happens on a very short time scale, typically in picoseconds. Hence, the effect of Akhiezer damping first becomes important at frequencies in the higher GHz regime.

In Fig. 3.15 the combined quality factor expected for thermoelastic and Akhiezer damping ($Q^{-1} = Q_{TED}^{-1} + Q_{Akhiezer}^{-1}$) of flexural beams of various materials is plotted as a function of the beam thickness h and vibrational frequency $\omega/(2\pi)$. It can be seen that the two fundamental loss mechanisms are strongly material dependent. On the one hand, thermoelastic damping in a typical nanomechanical beam ($10 < h < 1000$ nm and $L/h > 100$) allows for very high quality factors of the fundamental mode of more than a million for silicon. For thick silicon nitride beams, on the other hand, thermoelastic damping can become the limiting loss mechanism, an effect that has been observed with singly clamped cantilevers [56]. As a general rule, the effect of thermoelastic damping can be circumvented by designing thin beams (absolute, $h < 1$ μm, and relative, $L/h > 100$). Phonon-phonon interaction loss (Akhiezer effect) first starts to play a role at rather high frequencies in the GHz regime, which can be reached with very thin ($h < 10$ nm) and short ($L/h < 100$) beams. Higher-order flexural modes increase the vibrational frequency which generally increases the effect of thermoelastic and Akhiezer damping.

3.3.3 Dissipation Dilution

In the preceding sections, it was shown that the quality factor is directly proportional to the energy that gets dissipated by intrinsic loss mechanisms (3.30). Dissipation dilution describes the mechanism by which the influence of energy dissipation on the quality factor is reduced by the presence of a quasi-lossless extra potential. Dissipation dilution was, to our knowledge, for the first time reported and modeled independently in different fields by different research groups around 1990.

The notion that a quasi-lossless potential can dilute intrinsic losses was first formulated by P. Saulson in 1990 in a paper studying noise mechanisms in pendulum suspensions for the test masses used in a gravitational-wave interferometers [62].

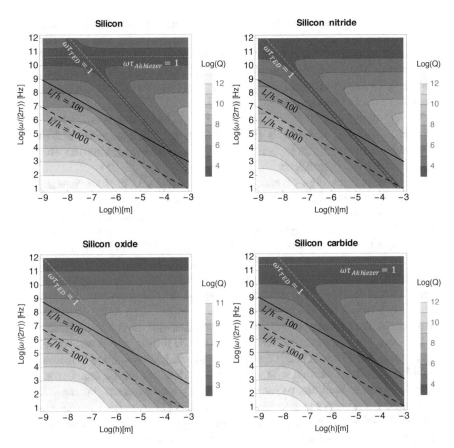

Fig. 3.15 Quality factors of vibrating beams due to thermoelastic and Akhiezer damping estimated for four typical nanomechanical materials as a function of beam thickness h and flexural frequency $\omega/(2\pi)$. The gray *dashed* and *dotted lines* represent the loss maxima for thermoelastic and Akhiezer losses. The black lines show the fundamental resonance frequencies of a doubly clamped beam resonator (bridge) of the specific material for two different ratios of length L to thickness h

In such a pendulum, the gravitational force pulling on a wire is free of loss. Saulson showed that the material loss in the wire gets diluted by a weighting factor that is given by the ratio between the energy stored in the flexing wire and the gravitational field. Since in a pendulum the latter is significantly larger, this scaling factor becomes much less than unity and hence he concluded: "Thus a pendulum can have much lower loss than the material of which it is made."

Contemporaneously, Berry et al. observed in 1989 inexplicably high quality factors in a few micrometer thin silicon as well as polymer membranes [63]. Shortly after, he presented an analytical theory explaining damping in stretched strings [64, 65]. In contrast to a pendulum, in strings and membranes, the lossless potential is presented by a static tensile stress instead of a gravitational force.

The dissipation dilution model for strings presented by Berry et al. already took nonlinear damping into account [64, 65]. However, the theory was incomplete by omitting to consider the dissipation associated with the local string bending at its clamping. The complete dissipation dilution model for a string resonator was derived by Gonzales and Saulson in 1994 [66]. It was also Saulson's group who later coined the term *dissipation dilution* [67].

Finally, dissipation dilution was described, modeled, and measured in 1993 for strings in music instruments [68]. Since music instruments are not played in vacuum, air damping is typically dominant and overshadowing the effect of dissipation dilution in instruments.

In the search for high-Q nanomechanical resonators, in 2006 inexplicably high quality factors were measured in stressed silicon nitride nanostrings [69]. A similar finding was made 2 years later with silicon nitride membranes, which showed quality factors of over one million at room temperature [70]. However, dissipation dilution theory was not recognized as an explanation for the high quality factors in nanomechanical silicon nitride resonators at the time. It was a study in 2008 of micromechanical polymer strings that linked the inexplicably high quality factors to dissipation dilution theory [19]. The high intrinsic friction loss in polymers makes the dissipation effect very obvious. In a sense, it has been a recurrence of the experience Berry et al. made when studying polymer membranes [63]. Dissipation dilution in nanomechanical silicon nitride strings was recognized shortly after [35, 71]. And in 2012, the theory for dissipation dilution in membranes was developed in connection with nanomechanical silicon nitride drumheads [72]. Recently, a generalized dissipation dilution theory was presented, linking it to the combination of tension and geometric nonlinearities [73].

In the derivation of the eigenfrequency of strings and membranes in Sects. 2.2.2 on page 43 and 2.3.2 on page 51, respectively, the assumption of a uniform static and constant tensile stress was made. It was further assumed that the flexural stiffness, and hence any bending moment, is so small that it can be neglected. These are valid assumptions when deriving the eigenfrequency. However, by neglecting extra elongation and curvature of a string or membrane structure during vibration, material losses are completely ignored, which would lead to the prediction of zero damping. Hence, in order to calculate the quality factor of strings and membranes, both the extensional and flexural strain energies have to be included. Accordingly, the quality factor (1.26 on page 9) of a string or a membrane can be expressed as [64, 65]

$$Q = 2\pi \frac{W_{tensile} + W_{elongation} + W_{bending}}{\Delta W_{elongation} + \Delta W_{bending}} \tag{3.49}$$

where $W_{tension}$ is the *lossless potential*, that is, the stored elastic energy required to deflect the string against the static tensile force. $\Delta W_{elongation}$ and $\Delta W_{bending}$, and $W_{elongation}$ and $W_{bending}$ are the lost and stored energy due to elongation and bending, respectively.

Assuming that the magnitude of the tensile pre-stress is dominating the mechanical behavior in strings and membranes, and by defining the intrinsic dissipation, as introduced in Sect. 3.3, to be equal for elongation and bending $Q_{intrinsic} = Q_{elongation} = \left(2\pi \frac{W_{elongation}}{\Delta W_{elongation}}\right) = Q_{bending} = \left(2\pi \frac{W_{bending}}{\Delta W_{bending}}\right)$, (3.49) can then be simplified to

$$Q \approx \alpha_{dd} \cdot Q_{intrinsic} \qquad (3.50)$$

with the *dissipation dilution factor*

$$\alpha_{dd} = \left[\frac{W_{bending}}{W_{tensile}} + \frac{W_{elongation}}{W_{tensile}}\right]^{-1}. \qquad (3.51)$$

Since, by definition, in a string or membrane the energy stored in the static tensile stress is a lot larger than the energies stored in bending and elongation $W_{tension} \gg W_{bending}, W_{elongation}$, the dissipation dilution factor becomes a lot larger than unity $\alpha_{dd} \gg 1$. This means that large potential energy from the tensile stress is "diluting" intrinsic losses and Q gets enhanced.

The effect of the additional energy stored in the tension described in (3.49 on the preceding page) can also be viewed from a different angle. According to (1.32 on page 10), the quality factor is the resonance frequency f_r divided by the peak width Δf:

$$Q = \frac{f_r}{\Delta f} = 2\pi \frac{W}{\Delta W}. \qquad (3.52)$$

Thus, the tensile tension increases the resonance frequency (which is equal to the stored energy) while the peak width (which is equal to the dissipation or energy lost) remains constant. Assuming that an added tensile stress does not alter the intrinsic losses ($\Delta f = $ constant), the tensile stress increases the frequency and hence the quality factor.

In the following, the dissipation dilution factor (3.51) will be derived for strings (Sect. 3.3.3.1) and membranes (Sect. 3.3.3.2. And in the last Sect. 3.3.3.3, so-called soft clamping will be briefly introduced, which is a way to enhance dissipation further and obtain ultrahigh quality factors.

3.3.3.1 Dissipation Dilution in Strings

The dissipation dilution in strings can readily be obtained from (3.51) based on the respective energies and the respective string modeshape given by (2.67)

$$U_n(x, t) = \phi_n(x)u_n(t) \qquad (3.53)$$

with the normalized mode shape

$$\phi_n(x) = \underbrace{\sin(\beta_\sigma x)}_{\text{sine shape}} + \underbrace{\frac{\beta_\sigma}{\beta_E} \left[e^{-\beta_E x} - \cos(\beta_\sigma x) \right]}_{\text{edge shape}} \tag{3.54}$$

and the wavenumbers related to the tensile stress $\beta_\sigma = \frac{n\pi}{L}$ and flexural rigidity of a membrane $\beta_E = \sqrt{\frac{\sigma A}{E I_y}}$, respectively.

The energy stored in the work against the tensile stress (see Table 2.1) can be calculated for the mode shape function (3.53):

$$W_{tension} = \frac{1}{2}\sigma A \int_0^L \left(\frac{\partial U_n}{\partial x} \right)^2 dx$$

$$\approx \frac{1}{4} u_n^2(t) \sigma A L \beta_\sigma^2. \tag{3.55}$$

Apparently, the tensile stored energy can be approximated by a single term coming from the bending of the sinusoidal mode shape. The effect of the exponential edge term in (3.53) is relatively small and can be dismissed for the calculation of the total tensile energy, as can be seen from the tensile energy density plotted in Fig. 3.16.

Similarly, the energy stored due to the string bending (from Table 2.1) can be calculated for the modeshape function (3.53)

$$W_{bending} = \frac{1}{2}E I_y \int_0^L \left(\frac{\partial^2 U_n}{\partial x^2} \right)^2 dx$$

$$\approx \underbrace{\frac{1}{4} u_n^2(t) E I_y \beta_\sigma^4 L}_{\text{sine shape}} + \underbrace{\frac{1}{2} u_n^2(t) E I_y \beta_\sigma^2 \beta_E}_{\text{edge shape}}, \tag{3.56}$$

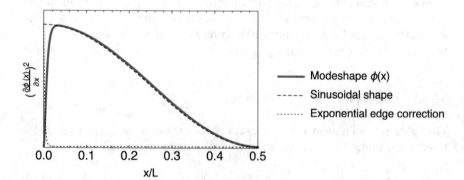

Fig. 3.16 Qualitative illustration of the tensile energy density in a string of length L

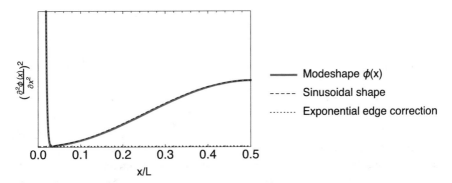

Fig. 3.17 Qualitative illustration of the bending energy density of a string of length L

which results in two terms: one coming from the antinodal bending of the sine mode shape and a second coming from the bending at the edge where the string is clamped. The bending energy density is illustrated in Fig. 3.17. From this plot it can be seen that the highest bending energy density is located near the edge of the string.

Equally, the energy stored in the elongation of the string (from Table 2.1) can be calculated for the modeshape function (3.53):

$$W_{elongation} = \frac{1}{8} E A \int_0^L \left(\frac{\partial U_n}{\partial x} \right)^4 dx$$

$$\approx \frac{3}{64} u_n^4(t) E A L \beta_\sigma^4.$$

(3.57)

Similar to the tensile energy, the elongational energy can be approximated by a single term coming from the sine modeshape. The effect of the exponential edge correction is insignificant and can be dismissed.

With (3.55), (3.56), and (3.57), the dissipation dilution factor (3.51) for the "one-dimensional case" of a string can now readily be calculated:

$$\alpha_{dd,1D} = \left[\underbrace{\left(\frac{\beta_\sigma}{\beta_E} \right)^2 + \frac{2}{\beta_E L}}_{\frac{W_{bending}}{W_{tensile}}} + \underbrace{\frac{9}{4} \left(\frac{u_n(t)}{h} \right)^2 \left(\frac{\beta_\sigma}{\beta_E} \right)^2}_{\frac{W_{elongation}}{W_{tensile}}} \right]^{-1},$$

(3.58)

which for a string with rectangular cross section with $I_y = \frac{wh^3}{12}$ becomes

$$
\alpha_{dd,1D} = \left[\underbrace{\underbrace{\frac{(n\pi)^2}{12} \frac{E}{\sigma} \left(\frac{h}{L}\right)^2}_{\text{sine shape}} + \underbrace{\frac{1}{\sqrt{3}} \sqrt{\frac{E}{\sigma}} \left(\frac{h}{L}\right)}_{\text{edge shape}}}_{\text{linear terms}} + \underbrace{\frac{3(n\pi)^2}{16} \frac{E}{\sigma} \left(\frac{u_n(t)}{L}\right)^2}_{\text{nonlinear term}} \right]^{-1} .
$$

(3.59)

The first two terms coming from the bending energy are the linear dissipation dilution terms and are independent of the vibrational amplitude $u_{0,n}$. In contrast, the term to the right, coming from the energy stored in the string elongation, is the nonlinear dissipation dilution term in a string. At large amplitudes, the elongational energy is accruing, and the measured quality factor becomes proportional to the vibrational amplitude squared, effectively resulting in nonlinear damping with the damping ratio $\zeta \approx 1/(2Q) \propto u_n^2(t)$. Such nonlinear damping can manifest in dissipation-diluted resonators when driven at large amplitudes. From the derivation of this nonlinear term in the dissipation dilution factor, it is clear that this effect of nonlinear damping has the same origin as the geometric nonlinearity as discussed in Sect. 2.7.1 on page 63. The string elongation becomes non-negligible at large vibrational amplitudes. It is important to understand that this nonlinear damping is a pure geometric and not a nonlinear dissipation effect, e.g., due to nonlinear material behavior.

Since for strings $h \ll L$, it is obvious that the edge shape term of the linear dissipation dilution in (3.59), representing the effect of the local bending at the clamping, is a lot larger than the sine shape term, representing the effect of the antinodal sine bending of the string. This means that energy loss near the edges is dominantly defining the quality factor in strings [35, 72]. The contribution of the energy loss due to the sinusoidal bending of the string (sine shape term) becomes only important for short strings at higher mode numbers. A schematic explanation of the linear dissipation dilution equation is shown in Fig. 3.18. And in Fig. 3.19 the linear dissipation dilution model is compared to measured Q values of nanomechanical silicon nitride string resonators. It shows how the quality factor decreases for increasing mode numbers and decreasing string length. The resulting intrinsic quality factor $Q_{intrinsic}$ from the comparison has been proposed to be due to surface losses as discussed in Sect. 3.3.1.2 on page 89 [36].

For small vibrational amplitudes $u_n(t) \ll h$, the nonlinear term is insignificant. In order for nonlinear damping to occur, the nonlinear term has to outgrow the dominating linear edge shape term that is coming from the dissipation in the local bending at the clamping. This typically is only fulfilled for rather large vibrational amplitudes $u_n(t) \gg h$. Additionally, higher harmonics and a low tensile stress can enhance the manifestation of nonlinear dissipation dilution. It is possible that conditions for nonlinear dissipation dilution can be found, e.g., in ultrathin carbon nanotube or graphene resonators. However, in so-called soft-clamped resonators, as discussed later in Sect. 3.3.3.3, any bending due to clamping disappears and with

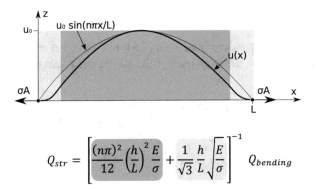

$$Q_{str} = \left[\frac{(n\pi)^2}{12} \left(\frac{h}{L}\right)^2 \frac{E}{\sigma} + \frac{1}{\sqrt{3}} \frac{h}{L} \sqrt{\frac{E}{\sigma}} \right]^{-1} Q_{bending}$$

Fig. 3.18 Schematic drawing explaining the dissipation dilution in strings (3.59) for generic bending-related damping mechanisms $Q_{intr} = Q_{bending}$. The first term comes from the damping due to the sinusoidal string bending. The second term comes from the local bending near the clamping

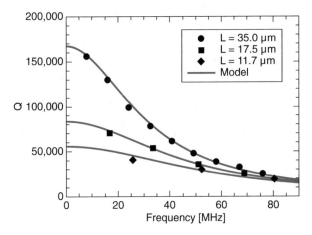

Fig. 3.19 Q values for increasing flexural modes of nanomechanical stoichiometric SiN strings with varying lengths ($\sigma = 942$ MPa, $h = 100$ nm) (data taken from [75]). All quality factors for the strings with different length are fitted with (3.59) with a single $Q_{intrinsic} = 4400$

it the otherwise dominating linear edge shape term in (3.59). In such soft-clamped resonators, the nonlinear term becomes significant already if $u_{0,n} \geq h$, as it has been observed with soft-clamped 2D drumheads made of silicon nitride [74].

3.3.3.2 Dissipation Dilution in Membranes

The dissipation dilution factor for a rectangular membrane with lateral dimensions L_x and L_y can readily be calculated from (3.51) with the membrane modeshape function:

$$U(x, y, t) = \phi_n(x)\phi_j(y)u(t) \tag{3.60}$$

with the normalized modeshape functions (3.54) in x- and y-direction.

The energy stored in the work against the tensile stress can be calculated from (see Table 2.1)

$$
\begin{aligned}
W_{tension} &= \frac{1}{2}\sigma h \int_0^{L_x}\int_0^{L_y}\left\{\left(\frac{\partial U_n}{\partial x}\right)^2 + \left(\frac{\partial U_n}{\partial y}\right)^2\right\}dxdy \\
&\approx \frac{1}{8}u_n^2(t)\sigma h L_x L_y\left(\beta_{\sigma,x}^2 + \beta_{\sigma,y}^2\right)
\end{aligned}
\tag{3.61}
$$

with the membrane wavenumbers $\beta_{\sigma,x} = \frac{n\pi}{L_x}$ and $\beta_{\sigma,y} = \frac{j\pi}{L_y}$. The energy stored due to the string bending can be calculated with (from Table 2.1)

$$
\begin{aligned}
W_{bending} &= \frac{D_P}{2}\int_0^{L_x}\int_0^{L_y}\left(\frac{\partial^2 U_n}{\partial x^2} + \frac{\partial^2 U_n}{\partial y^2}\right)^2 \\
&\quad + 2(1-v)\left[\left(\frac{\partial^2 U}{\partial x\partial y}\right)^2 - \frac{\partial^2 U_n}{\partial x^2}\frac{\partial^2 U_n}{\partial y^2}\right]dxdy
\end{aligned}
\tag{3.62}
$$

with the flexural rigidity $D_P = Eh^3/(12(1-v^2))$. For a homogeneous membrane with constant thickness clamped at all four boundaries, the second term in (3.62) becomes zero according to Green's theorem [72]. The bending energy can thus be calculated by the simplified form:

$$
\begin{aligned}
W_{bending} &= \frac{D_P}{2}\int_0^{L_x}\int_0^{L_y}\left(\frac{\partial^2 U_n}{\partial x^2} + \frac{\partial^2 U_n}{\partial y^2}\right)^2 dxdy \\
&\approx \underbrace{\frac{1}{8}u_n^2(t)D_P L_x L_y\left(\beta_{\sigma,x}^2 + \beta_{\sigma,y}\right)^2}_{\text{sine shape}} + \underbrace{\frac{1}{4}u_n^2(t)D_P\beta_E\left(L_x\beta_{\sigma,y}^2 + L_y\beta_{\sigma,x}^2\right)}_{\text{edge shape}},
\end{aligned}
\tag{3.63}
$$

with the flexural wavenumber $\beta_E = \sqrt{\frac{\sigma h}{D_P}}$. The bending energy in a membrane, similar to the string case presented in the previous subsection, results in two terms: one coming from the antinodal bending of the sine modeshape and a second coming from the bending at the edge where the membrane is clamped.

For small vibrational amplitudes, the "two-dimensional" dissipation dilution factor for a rectangular membrane then becomes:

$$\alpha_{dd,2D} \approx \left[\underbrace{\frac{\left(\beta_{\sigma,x}^2 + \beta_{\sigma,y}^2\right)}{\beta_E^2}}_{\text{sine shape}} + \underbrace{\frac{2}{\beta_E} \frac{\left(L_x\beta_{\sigma,y}^2 + L_y\beta_{\sigma,x}^2\right)}{L_x L_y \left(\beta_{\sigma,x}^2 + \beta_{\sigma,y}^2\right)}}_{\text{edge shape}} \right]^{-1}$$

$$\approx \left[\frac{D_P}{\sigma h} \pi^2 \left(\left(\frac{n}{L_x}\right)^2 + \left(\frac{j}{L_y}\right)^2 \right) + 2\sqrt{\frac{D_P}{\sigma h}} \frac{L_y\left(\frac{n}{L_x}\right)^2 + L_x\left(\frac{j}{L_y}\right)^2}{L_x L_y \left(\left(\frac{n}{L_x}\right)^2 + \left(\frac{j}{L_y}\right)^2 \right)} \right]^{-1}$$

$$(3.64)$$

which can further be simplified for a square membrane ($L_x = L_y = L$) and assuming zero transversal strain ($\nu = 0$)

$$\alpha_{dd,2D} \approx \left[\underbrace{\frac{\pi^2(n^2 + j^2)}{12} \frac{E}{\sigma} \left(\frac{h}{L}\right)^2}_{\text{sine shape}} + \underbrace{\frac{1}{\sqrt{3}} \sqrt{\frac{E}{\sigma}} \left(\frac{h}{L}\right)}_{\text{edge shape}} \right]^{-1} . \qquad (3.65)$$

The resulting dissipation dilution factor for a square membrane is very similar to the factor for a string with rectangular cross section (3.59).

Calculating a closed form solution of the nonlinear term of the dissipation dilution factor for a 2D resonator is challenging because of the non-negligible lateral displacement field [76]. A general solution can be found in Catalini et al. [74].

3.3.3.3 Soft Clamping

The term *soft clamping* was coined by Tsaturian et al. in their seminal paper in 2017, where they demonstrated a localized bending defect mode inside a phononic crystal structure that was patterned on a high-stress silicon nitride membrane [48]. Such an exemplary structure is shown in Fig. 3.20a. The phononic crystal has two effects. First, the phononic crystal produced a quasi-band gap featuring a localized mode in the center defect as a way to eliminate clamping losses, as discussed previously in Sect. 3.2.3 on page 80. Second, and even more interesting, the resulting evanescent (soft) decay of the localized mode into the phononic crystal eliminates the loss-related strong curvature that would occur in a clamped resonator of similar effective size. This effect is schematically presented in Fig. 3.20b. Soft clamping effectively eliminates the dominating edge shape term in the dissipation dilution factor: (3.65)

$$\alpha_{dd,2D\text{-}SC} \approx \frac{12}{\pi^2(n^2 + j^2)} \frac{\sigma}{E} \left(\frac{L}{h}\right)^2 . \qquad (3.66)$$

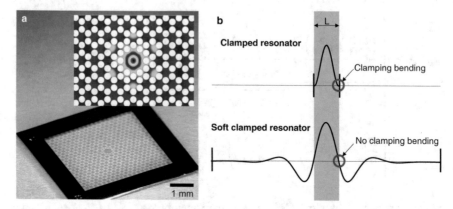

Fig. 3.20 (**a**) Photograph of a LPCVD silicon nitride membrane featuring with a phononic crystal pattern featuring a defect in the center. The inset shows an FEM simulation of the mechanical defect mode. (**b**) A schematic representation of the mode shape of a regularly clamped resonator and a soft clamped resonator

Soft clamping has resulted in an quality factor enhancement of two orders of magnitude, compared to a drumhead resonator of similar effective lateral size L, reaching values of $Q > 10^8$ at a resonance frequency of 1 MHz [48]. This specific phononic crystal design has shown to be insensitive to the value of the tensile stress in the supporting SiN membrane [77]. Other phononic crystal design variations have also been successfully demonstrated and analyzed for their soft clamping characteristics [78].

Besides 2D phononic crystals, soft clamping has also been demonstrated with 1D phononic crystals pattern on a high-stress silicon nitride string by Ghadimi et al. [49]. In addition to the soft clamping, the tensile stress in the center defect was enhanced by tapering of the string width toward the localized defect mode. The combination of soft clamping and stress tuning resulted in quality factors $Q > 8 \times 10^8$ at 1.3 MHz. Fractal-like designs for soft clamping of string resonators have been presented as a design alternative to phononic crystals [79].

In addition to the use of phononic crystals, there are other approaches to achieve soft clamping. It has been shown that perimeter modes in rectangular or polygon-shaped string resonators that are tethered at its corners produce a soft clamping resulting in Q values of several billions [80, 81]. Also, dissipation dilution optimization strategies based on optimized clamping design in trampoline and string resonators have been considered. It was shown that clamp widening (reducing the local stress at the clamp) can't significantly reduce the local curvature at the clamping points [82]. But it has been demonstrated that tapering of a string near the clamping (increasing the local stress at the clamp) can improve dissipation dilution to some degree [83]. An improvement of Q in SiN trampoline resonators has been achieved by topology optimization of the resonator geometry [84].

References

1. Stipe, B. C., Mamin, H. J., Stowe, T. D., Kenny, T. W., & Rugar, D. (2001). Noncontact friction and force fluctuations between closely spaced bodies. *Physical Review Letters, 87*(9), 096801.
2. Cleland, A. N., & Roukes, M. L. (1999). External control of dissipation in a nanometer-scale radiofrequency mechanical resonator. *Sensors and Actuators A: Physical, 72*(3), 256–261.
3. Schwab, K. (2002). Spring constant and damping constant tuning of nanomechanical resonators using a single-electron transistor. *Applied Physics Letters, 80*(7), 1276–1278.
4. Sader, J. E. (1998). Frequency response of cantilever beams immersed in viscous fluids with applications to the atomic force microscope. *Journal of Applied Physics, 84*(1), 64.
5. Van Eysden, C. a., & Sader, J. E. (2007). Frequency response of cantilever beams immersed in viscous fluids with applications to the atomic force microscope: Arbitrary mode order. *Journal of Applied Physics, 101*(4), 044908.
6. Green, C. P., & Sader, J. E. (2002). Torsional frequency response of cantilever beams immersed in viscous fluids with applications to the atomic force microscope. *Journal of Applied Physics, 92*(10), 6262.
7. Green, C. P., & Sader, J. E. (2005). Frequency response of cantilever beams immersed in viscous fluids near a solid surface with applications to the atomic force microscope. *Journal of Applied Physics, 98*(11), 114913.
8. Ghatkesar, M. K., Braun, T., Barwich, V., Ramseyer, J.-P., Gerber, C., Hegner, M., & Lang, H. P. (2008). Resonating modes of vibrating microcantilevers in liquid. *Applied Physics Letters, 92*(4), 043106.
9. Chon, J. W. M., Mulvaney, P., & Sader, J. E. (2000). Experimental validation of theoretical models for the frequency response of atomic force microscope cantilever beams immersed in fluids. *Journal of Applied Physics, 87*(8), 3978.
10. Van Eysden, C. a., & Sader, J. E. (2006). Resonant frequencies of a rectangular cantilever beam immersed in a fluid. *Journal of Applied Physics, 100*(11), 114916.
11. Burg, T. P., Godin, M., Knudsen, S. M., Shen, W., Carlson, G., Foster, J. S., Babcock, K., & Manalis, S. R. (2007). Weighing of biomolecules, single cells and single nanoparticles in fluid. *Nature, 446*(7139), 1066–1069.
12. Lee, J., Shen, W., Payer, K., Burg, T. P., & Manalis, S. R. (2010). Toward attogram mass measurements in solution with suspended nanochannel resonators. *Nano Letters, 10*(7), 2537–2542.
13. Enoksson, P., Stemme, G., & Stemme, E. (1995). Fluid density sensor based on resonance vibration. *Sensors and Actuators A: Physical, 47*, 327–331.
14. Sparks, D., Schneider, R., Smith, R., Chimbayo, A., Straayer, M., Cripe, J., Najafi, N., & Anasari, S. (2003). Measurement of density and chemical concentration using a microfluidic chip. *Lab on a Chip, 3*(1), 19–21.
15. Westberg, D., Paul, O., Andersson, G. I., & Baltes, H. (1999). A CMOS-compatible device for fluid density measurements fabricated by sacrificial aluminium etching. *Sensors and Actuators A: Physical, 73*(3), 243–251.
16. Khan, M. F., Schmid, S., Larsen, P. E., Davis, Z. J., Yan, W., Stenby, E. H., & Boisen, a. (2013). Online measurement of mass density and viscosity of pL fluid samples with suspended microchannel resonator. *Sensors and Actuators B: Chemical, 185*, 456–461.
17. Burg, T., Sader, J., & Manalis, S. (2009). Nonmonotonic energy dissipation in microfluidic resonators. *Physical Review Letters, 102*(22), 1–4.
18. Sader, J. E., Burg, T. P., & Manalis, S. R. (2010). Energy dissipation in microfluidic beam resonators. *Journal of Fluid Mechanics, 650*, 215–250.
19. Schmid, S., & Hierold, C. (2008). Damping mechanisms of single-clamped and prestressed double-clamped resonant polymer microbeams. *Journal of Applied Physics, 104*(9), 093516.
20. Van Eysden, C. A., & Sader, J. E. (2009). Frequency response of cantilever beams immersed in compressible fluids with applications to the atomic force microscope. *Journal of Applied Physics, 106*(9), 094904.

21. Bao, M. (2005). *Analysis and design principles of MEMS devices*. Amsterdam: Elsevier.
22. Verbridge, S. S., Ilic, R., Craighead, H. G., & Parpia, J. M. (2008). Size and frequency dependent gas damping of nanomechanical resonators. *Applied Physics Letters, 93*(1), 13101.
23. Schmid, S., Malm, B., & Boisen, A. (2011). Quality factor improvement of silicon nitride micro string resonators. In *24th international conference on Micro Electro Mechanical Systems (MEMS)* (pp. 481–484). New York: IEEE.
24. Bao, M., Yang, H., Yin, H., & Sun, Y. (2002). Energy transfer model for squeeze-film air damping in low vacuum. *Journal of Micromechanics and Microengineering, 12*, 341–346.
25. Christian, R. G. (1966). The theory of oscillating-vane vacuum gauges. *Vacuum, 16*, 175–178.
26. Li, P., & Hu, R. (2007). On the air damping of flexible microbeam in free space at the free-molecule regime. *Microfluidics and Nanofluidics, 3*, 715–721.
27. Photiadis, D. M., & Judge, J. A. (2004). Attachment losses of high Q oscillators. *Applied Physics Letters, 85*(3), 482–484.
28. Cross, M. C., & Lifshitz, R. (2001). Elastic wave transmission at an abrupt junction in a thin plate with application to heat transport and vibrations in mesoscopic systems. *Physical Review B, 64*(8), 85324.
29. Schmid, S. (2009). Electrostatically actuated all-polymer microbeam resonators—characterization and application. In *Scientific reports on micro and nanosystems* (Vol. 6). New York: Der Andere Verlag.
30. Wilson-Rae, I. (2008). Intrinsic dissipation in nanomechanical resonators due to phonon tunneling. *Physical Review B, 77*(24), 245418.
31. Wilson-Rae, I., Barton, R. A., Verbridge, S. S., Southworth, D. R., Ilic, B., Craighead, H. G., & Parpia, J. M. (2011). High-Q nanomechanics via destructive interference of elastic waves. *Physical Review Letters, 106*(4), 47205.
32. Wilson, D. J., Regal, C. A., Papp, S. B., & Kimble, H. J. (2009). Cavity optomechanics with stoichiometric SiN films. *Physical Review Letters, 103*(20), 207204.
33. Wilson, D. J. (2012). *Cavity optomechanics with high-stress silicon nitride films*. PhD thesis, California: California Institute of Technology.
34. Chakram, S., Patil, Y. S., Chang, L., & Vengalattore, M. (2014). Dissipation in ultrahigh quality factor SiN membrane resonators. *Physical Review Letters, 112*(12), 127201.
35. Schmid, S., Jensen, K. D., Nielsen, K. H., & Boisen, A. (2011). Damping mechanisms in high-Q micro and nanomechanical string resonators. *Physical Review B, 84*(16), 165307.
36. Villanueva, L. G., & Schmid, S. (2014). Evidence of surface loss as ubiquitous limiting damping mechanism in SiN micro- and nanomechanical resonators. *Physical Review Letters, 113*(227201), 1–6.
37. Lozzi, A., De Pastina, A., Yen, E. T. T., & Villanueva, L. G. (2019). Engineered acoustic mismatch for anchor loss control in contour mode resonators. *Applied Physics Letters, 114*(10), 103502.
38. Hales, J. H., Teva, J., Boisen, A., & Davis, Z. J. (2009). Longitudinal bulk acoustic mass sensor. *Applied Physics Letters, 95*(3), 033506.
39. Wang, K., Wong, A.-C., & Nguyen, C. T. (2000). Vhf free-free beam high-q micromechanical resonators. *Journal of Microelectromechanical Systems, 9*(3), 347–360
40. Tsaturyan, Y., Barg, A., Simonsen, A., Villanueva, L. G., Schmid, S., Schliesser, A., & Polzik, E. S. (2014). Demonstration of suppressed phonon tunneling losses in phononic bandgap shielded membrane resonators for high-Q optomechanics. *Optics Express, 22*(6), 6810.
41. Yu, P.-L., Cicak, K., Kampel, N. S., Tsaturyan, Y., Purdy, T. P., Simmonds, R. W., & Regal, C. A. (2014). A phononic bandgap shield for high-Q membrane microresonators. *Applied Physics Letters, 104*(2), 023510.
42. Hsu, F.-C., Hsu, J.-C., Huang, T.-C., Wang, C.-H., & Chang, P. (2011). Design of lossless anchors for microacoustic-wave resonators utilizing phononic crystal strips. *Applied Physics Letters, 98*(14), 143505.
43. Mohammadi, S., Eftekhar, A. A., Hunt, W. D., & Adibi, A. (2009). High-q micromechanical resonators in a two-dimensional phononic crystal slab. *Applied Physics Letters, 94*(5), 051906.

44. Mohammadi, S., & Adibi, A. (2012). Waveguide-based phononic crystal micro/nanomechanical high-q resonators. *Journal of Microelectromechanical Systems, 21*(2), 379–384.

45. Eichenfield, M., Chan, J., Camacho, R. M., Vahala, K. J., & Painter, O. (2009). Optomechanical crystals. *Nature, 462*(7269), 78–82.

46. Chan, J., Alegre, T. P. M., Safavi-Naeini, A. H., Hill, J. T., Krause, A., Gröblacher, S., Aspelmeyer, M., & Painter, O. (2011). Laser cooling of a nanomechanical oscillator into its quantum ground state. *Nature, 478*(7367), 89–92.

47. Maldovan, M., & Thomas, E. L. (2006). Simultaneous localization of photons and phonons in two-dimensional periodic structures. *Applied Physics Letters, 88*(25), 20–23.

48. Tsaturyan, Y., Barg, A., Polzik, E. S., & Schliesser, A. (2017). Ultracoherent nanomechanical resonators via soft clamping and dissipation dilution. *Nature Nanotechnology, 12*(8), 776.

49. Ghadimi, A. H., Fedorov, S. A., Engelsen, N. J., Bereyhi, M. J., Schilling, R., Wilson, D. J., & Kippenberg, T. J. (2018). Elastic strain engineering for ultralow mechanical dissipation. *Science, 360*(6390), 764–768.

50. Ashby, M. F., & Ashby, F. (1989). Overview No. 80: On the engineering properties of materials. *Acta Metallurgica, 37*(5), 1273–1293.

51. Fantozzi, G. (2001). 1.1 Phenomenology and Definitions. *Materials Science Forum, 366–368*, 3–31.

52. Ward, I. M., & Sweeney, J. (2004). *An introduction to the mechanical properties of solid polymers* (2nd ed.). London: Wiley.

53. Yang, J., Ono, T., & Esashi, M. (2002). Energy dissipation in submicrometer thick single-crystal cantilevers. *Journal of Microelectromechanical Systems, 11*(6), 775–783.

54. Mohanty, P., Harrington, D. A., Ekinci, K. L., Yang, Y. T., Murphy, M. J., & Roukes, M. L. (2002). Intrinsic dissipation in high-frequency micromechanical resonators. *Physical Review B, 66*(8), 85416.

55. Tao, Y., Navaretti, P., Hauert, R., Grob, U., Poggio, M., & Degen, C. L. (2015). Permanent reduction of dissipation in nanomechanical Si resonators by chemical surface protection. *Nanotechnology, 26*(46), 465501.

56. Yasumura, K. Y., Stowe, T. D., Chow, E. M., Pfafman, T., Kenny, T. W., Stipe, B. C., & Rugar, D. (2000). Quality factors in micron- and submicron-thick cantilevers. *Journal of Microelectromechanical Systems, 9*(1), 117–125.

57. Cleland, A. N. (2003). *Foundations of nanomechanics.* New York: Springer.

58. Roszhart, T. V. (1990). The effect of thermoelastic internal friction on the Q of micromachined silicon resonators. In *IEEE 4th technical digest on solid-state sensor and actuator workshop* (pp. 13–16).

59. Zener, C. (1937). Internal friction in solids {I}: {T}heory of internal friction in reeds. *Physical Review, 52*, 230–235.

60. Lifshitz, R., & Roukes, M. L. (2000). Thermoelastic damping in micro- and nanomechanical systems. *Physical Review B, 61*(8), 5600–5609.

61. Prabhakar, S., & Vengallatore, S. (2007). Thermoelastic damping in bilayered micromechanical beam resonators. *Journal of Micromechanics and Microengineering, 17*(3), 532–538.

62. Saulson, P. R. (1990). Thermal noise in mechanical experiments. *Physical Review D, 42*, 2437–2445.

63. Berry, B. S., Pritchet, W. C., & Uzoh, C. E. (1989). Dynamical method for the thermomechanical study of thin membranes. *Journal of Vacuum Science and Technology B: Microelectronics Processing and Phenomena, 7*(6), 1565–1569.

64. Berry, B. S., & Pritchet, W. C. (1991). New developments in the dynamic mechanical analysis of thin-layer materials. In *MRS online proceedings library archive* (p. 226).

65. Berry, B. S. (1992). Damping mechanisms in thin-layer materials. In *M 3 D: Mechanics and Mechanisms of Material Damping.* New York: ASTM International.

66. Gonzfilez, G. I., & Saulson, P. R. (1994). Brownian motion of a mass suspended by an anelastic wire. *Journal of the Acoustical Society of America, 96*(1), 207–212.

67. Huang, Y. L., & Saulson, P. R. (1998). Dissipation mechanisms in pendulums and their implications for gravitational wave interferometers. *Review of Scientific Instruments, 69*(2), 544.
68. Valette, C., & Cuesta, C. (1993). *Mécanique de la corde vibrante*. Paris: Hermes.
69. Verbridge, S. S., Parpia, J. M., Reichenbach, R. B., Bellan, L. M., & Craighead, H. G. (2006). High quality factor resonance at room temperature with nanostrings under high tensile stress. *Journal of Applied Physics, 99*, 124304.
70. Zwickl, B. M., Shanks, W. E., Jayich, A. M., Yang, C., Jayich, B., Thompson, J. D., & Harris, J. G. E. (2008). High quality mechanical and optical properties of commercial silicon nitride membranes. *Applied Physics Letters, 92*(10), 103125.
71. Unterreithmeier, Q. P., Faust, T., & Kotthaus, J. P. (2010). Damping of nanomechanical resonators. *Physical Review Letters, 105*, 027205.
72. Yu, P.-L., Purdy, T., & Regal, C. A. (2012). Control of material damping in high-Q membrane microresonators. *Physical Review Letters, 108*(8), 083603.
73. Fedorov, S. A., Engelsen, N. J., Ghadimi, A. H., Bereyhi, M. J., Schilling, R., Wilson, D. J., & Kippenberg, T. J. (2019). Generalized dissipation dilution in strained mechanical resonators. *Physical Review B, 99*(5), 054107.
74. Catalini, L., Rossi, M., Langman, E. C., & Schliesser, A. (2021). Modeling and observation of nonlinear damping in dissipation-diluted nanomechanical resonators. *Physical Review Letters, 126*(17), 174101.
75. Unterreithmeier, Q. P., Faust, T., & Kotthaus, J. P. (2010). Damping of nanomechanical resonators. *Physical Review Letters, 105*(2), 27205.
76. Ventsel, E., & Krauthammer, T. (2001). *Thin plates and shells: Theory, analysis, and applications*. New York: Marcel Dekker.
77. Sadeghi, P., Tanzer, M., Luhmann, N., Piller, M., Chien, M. H., & Schmid, S. (2020). Thermal transport and frequency response of localized modes on low-stress nanomechanical silicon nitride drums featuring a phononic-band-gap structure. *Physical Review Applied, 14*(2), 024068.
78. Reetz, C., Fischer, R., Assumpcao, G. G., McNally, D. P., Burns, P. S., Sankey, J. C., & Regal, C. A. (2019). Analysis of membrane phononic crystals with wide band gaps and low-mass defects. *Physical Review Applied, 12*(4), 044027.
79. Fedorov, S. A., Beccari, A., Engelsen, N. J., & Kippenberg, T. J. (2020). Fractal-like mechanical resonators with a soft-clamped fundamental mode. *Physical Review Letters, 124*, 025502.
80. Shin, D., Cupertino, A., de Jong, M. H., Steeneken, P. G., Bessa, M. A., & Norte, R. A. (2022). Spiderweb nanomechanical resonators via bayesian optimization: Inspired by nature and guided by machine learning. *Advanced Materials, 34*(3), 2106248.
81. Bereyhi, M., Arabmoheghi, A., Fedorov, S. A., Beccari, A., Huang, G., Kippenberg, T. J., Engelsen, N. J., et al. (2021). Nanomechanical resonators with ultra-high-*q* perimeter modes. *arXiv preprint arXiv:2108.03615*.
82. Sadeghi, P., Tanzer, M., Christensen, S. L., & Schmid, S. (2019). Influence of clamp-widening on the quality factor of nanomechanical silicon nitride resonators. *Journal of Applied Physics, 126*(16), 165108.
83. Bereyhi, M. J., Beccari, A., Fedorov, S. A., Ghadimi, A. H., Schilling, R., Wilson, D. J., Engelsen, N. J., & Kippenberg, T. J. (2019). Clamp-tapering increases the quality factor of stressed nanobeams. *Nano Letters, 19*(4), 2329–2333.
84. Høj, D., Wang, F., Gao, W., Hoff, U. B., Sigmund, O., & Andersen, U. L. (2021). Ultra-coherent nanomechanical resonators based on inverse design. *arXiv preprint arXiv:2103.15601*.

Chapter 4
Transduction

This is an overview of techniques commonly used to transduce (actuate and detect) the mechanical motion of nanomechanical resonators. Every technique is briefly explained and prominent corresponding examples are presented. The reference list is not complete by any means. The idea of this overview is to call the reader's attention to all possible transduction techniques and give him/her an entry point for his/her own thorough literature research.

There are a few techniques that are less common, which in this edition will be omitted. This, e.g., includes the detection of motion of nanomechanical resonators by tunneling [1], hard contact [2], or field emission[3]. Another technique which will not be discussed in more detail is *magnetostatic transduction*. This includes the transduction of ferromagnetic [4] or paramagnetic [5] mechanical resonators with the help of an external dynamic magnetic field. The increasing inductive reactance of a coil with increasing frequency sets a limit at room temperature of the maximal achievable frequency in the lower kHz range. Hence, this technique is rarely used to transduce nanomechanical resonators with frequencies in the MHz range.

The focus of this book is NEMS that is nanomechanical resonators that are transduced by electrical means. Besides electrical transduction, optical detection of nanomechanical motion is commonly used, which can be external or fully integrated. Like many electrical transduction techniques, optical transduction can be used not only to detect but also to actuate a nanomechanical resonator by means of radiation pressure. The full optical transduction and control of nanomechanical resonators is studied in cavity optomechanics [6, 7]. Cavity optomechanics has developed into a field, and its specific transduction methods are not covered in this chapter.

Some transduction techniques allow for both actuation and detection, whereas others only work for either detection or actuation. The nature of each technique is assigned in the parentheses behind the title.

© The Author(s), under exclusive license to Springer Nature Switzerland AG 2023
S. Schmid et al., *Fundamentals of Nanomechanical Resonators*,
https://doi.org/10.1007/978-3-031-29628-4_4

4.1   Electrodynamic Actuation and Detection

Electrodynamic transduction is based on the *Lorentz force* acting on free charges inside a wire that is located in a uniform magnetic field **B**. The Lorentz force **F** acting on a single particle with charge q in the absence of an electric field is then given by

$$\mathbf{F_L} = q\mathbf{v} \times \mathbf{B} \tag{4.1}$$

with the velocity vector **v**. From the vector product, it is clear that the resulting force is acting in the direction normal to the plane spanned by the magnetic field and velocity vector. The effect of the Lorentz force can be used to actuate and detect the motion of a nanomechanical resonator, as discussed subsequently.

Electrodynamic transduction is relatively easy to implement and allows for an efficient actuation and detection of nanomechanical resonators with frequencies up to the GHz range. It is thus not surprising that electrodynamic transduction had been an enabling technique for the pioneering examples of nanomechanical resonators [8] (see Fig. 4.1).

4.1.1   Lorentz Force on a Wire

A conductive wire carrying an electric current experiences the Lorentz force when it is located inside a uniform magnetic field **B**. Individual electric charges that are traveling with a velocity **v** inside the wire each experience the Lorentz force (4.1). The sum of all the forces acting on all individual charges creates a macroscopic force on the wire. The electric current inside a wire of length L is defined by

$$\mathbf{I} = \frac{nq\mathbf{v}}{L} \tag{4.2}$$

Fig. 4.1 SEM micrograph of a nanomechanical silicon resonator fully transduced electrodynamically [8]

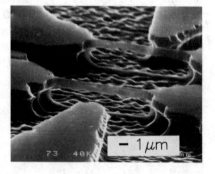

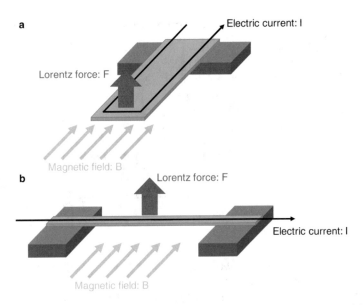

Fig. 4.2 Schematic drawings of two common electrode configuration to actuate (**a**) singly clamped and (**b**) doubly clamped beams

with the total number of charges n. Combining this definition with the Lorentz force law (4.1) yields a term for the total magnetic force acting on the wire as a function of the electric current:

$$\mathbf{F} = L\mathbf{I} \times \mathbf{B}. \tag{4.3}$$

The most common ways to implement a Lorentz force actuation creating an out-of-plane vibration in singly clamped and doubly clamped nanomechanical beam resonators are shown in Fig. 4.2b. Assuming a normalized sinusoidal modeshape function (2.49 on page 43)

$$\phi(x) = \sin\left(\frac{n\pi}{L}x\right), \tag{4.4}$$

the effective force (2.130 on page 62) acting on such a string can be calculated by

$$
\begin{aligned}
F_{eff,n} &= IB \int_0^L \phi(x)\mathrm{d}x \\
&= IBL\frac{1 - \cos(n\pi)}{n\pi}.
\end{aligned}
\tag{4.5}
$$

The equation above implies that the effective force is zero for even mode numbers n. As a result, only odd resonant bending modes can be actuated with this technique.

Fig. 4.3 Frequency response
of a gold-coated silicon
nitride microstring (340 nm
thick and 1000 μm long)
placed in a magnetic Halbach
array with a field strength of
$B = 0.4$ T and actuated with
an ac voltage of 30 mV. The
resonance was detected with
a laser-Doppler vibrometer
(MSA-500 from Polytec
GmbH) at atmospheric
pressure

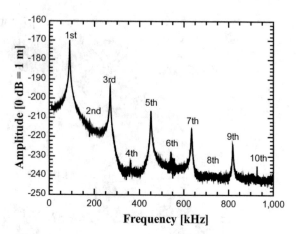

This effect can be seen in the frequency response of a resonant string actuated by
the Lorentz force shown in Fig. 4.3.

The Lorentz force (4.3) is increasing linearly with the magnetic field strength. At
room temperature a strong magnetic field can be achieved with rare earth magnets.
When arranging them in a Halbach array, magnetic field strengths of up to $B = 2$ T
can be achieved [9]. At cryogenic temperatures, magnetic fields of up to $B = 7-8$ T
can be achieved with superconducting coils [8, 10].

4.1.2 Electrodynamically Induced Voltage (Electromotive Force)

When a wire is moving inside a uniform magnetic field, the individual free
charges in the wire experience the Lorentz force (4.1). In the case of the depicted
transduction schemes in Fig. 4.2, the magnetic forces are pushing the charges in the
direction of the wire. This magnetic force acting on the charges results in a charge
separation and hence a potential difference across the length of the wire. In the
case of a moving conducting wire, this voltage is called the motional *electromotive
force* (V_{EMF}). This induced voltage is the underlying principle of typical electric
generators.

The potential difference, or voltage, between two points is defined as the work
done per unit charge to move the charge from one point to the other. Hence, with
the Lorentz force (4.1) acting on a single charge, the EMF between the ends of a
straight wire element of length dx is given by

$$dV_{EMF} = \frac{F_L}{q}dx = Bvdx. \tag{4.6}$$

Considering the transduction scheme depicted in Fig. 4.2, the velocity of a specific wire element depends on its position along the length of the string with length L. It is then straightforward to calculate the potential difference that can be picked up at the ends of such a wire that is vibrating in a static magnetic field by integrating the partial potential differences for all string elements (4.6). Assuming an ideal sinusoidal modeshape function (2.49 on page 43)

$$U(x,t) = u_0 \sin\left(\frac{n\pi}{L}x\right)\cos\omega_n t, \tag{4.7}$$

the total EMF voltage resulting at the ends of a vibrating string resonator can be calculated by

$$\begin{aligned} V_{EMF}(t) &= \int_0^L dV_{EMF} \\ &= B \int_0^L \frac{\partial U(x,t)}{\partial t} dx. \end{aligned} \tag{4.8}$$

This results in a root mean square voltage:

$$V_{EMF_{RMS}} = B\sqrt{2}u_0 f_1 L\left[1 - \cos(n\pi)\right] \tag{4.9}$$

with $\omega_n = 2\pi f_n$ and $f_1 = f_n/n$ (2.55 on page 44). The above equation implies that the voltage is zero for even mode numbers n, as already discussed for the actuation previously.

The detection of this EMF is straightforward, and it can directly be picked up, e.g., with a network analyzer or a lock-in amplifier [8, 9], as schematically depicted in Fig. 4.4. More complex detection schemes consist of differential setup with a passive reference device for the use in a self-sustaining oscillator circuit[10].

While this transduction technique is simple, there is a caveat. The magnetomotively induced current, induced by (4.6), has a 90-degree phase offset with respect

Fig. 4.4 Schematic of simple electrodynamic transduction including actuation by the Lorentz force detection of electromotive current

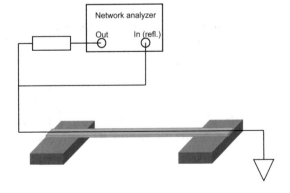

to the vibrating oscillation, which can cause significant magnetomotive damping of the mechanical vibration [11, 12]. Magnetomotive damping can be minimized by reducing the current, which can be achieved, e.g., by introducing a large resistance in series with the transducing electrodes.

4.2   Electrostatic Actuation and Detection

Electrostatic transduction allows for a low-power operation and uncomplicated system integration. Therefore, capacitive transduction schemes have become a cornerstone in microelectromechanical systems (MEMS), where they are very successfully used in devices such as accelerometers, gyroscopes, microphones, pressure sensors, and mirror arrays for projectors, etc. The low-power consumption is equally interesting for the use in smaller nanomechanical resonators. However, the smaller size of the moving parts in nano-sized resonators results in smaller electrostatic signals. In this subsection electrostatic transduction techniques are discussed for the application in nanomechanical resonators.

Generally, electrostatic transduction is based on the interaction between quasi-static electric charges. Practically, the electric charge in a nanomechanical resonator is controlled via conductive electrodes with a defined potential. This scheme of inducing charges by means of a defined potential is the main scope of this subsection. In contrast, electric charges can also locally be trapped, typically at the interface of different materials, in particular on dielectrics. It has, e.g., been shown that local charge separation, resulting in a static dipole moment, can be used to transduce nanomechanical resonators made of GaAs/AlGaAs multilayers[13]. While local charges can be a key part of a transducer design, e.g., in electret microphones, locally trapped charges can hinder an effective electrostatic transduction [14].

4.2.1   Electrostatic Forces

In a lossless system, the force in direction of a specific degree of freedom ξ is given by the change of potential energy W_e of the system in the ξ-direction:

$$F = -\frac{\partial W_e(\xi)}{\partial \xi}. \tag{4.10}$$

The energy W_e stored in an electrostatic system can be derived by integration of the energy density over the significant system volume v:

$$W_e = \int_v \frac{1}{2}\varepsilon_r\varepsilon_0 E^2 dv \tag{4.11}$$

where E is the electrostatic field strength, ε_r is the relative dielectric constant, and ε_0 is the permittivity of vacuum. In an electrostatic system with a single electrical terminal pair with a constant potential V, the capacity of the system to store electrostatic energy as a function of the applied potential is typically given as a function of the systems capacitance C, given by

$$W_e = -\frac{1}{2}CV^2. \tag{4.12}$$

Assuming that the system has a single degree of freedom ξ, the electrostatic force in ξ-direction is then, according to (4.10), given by the change of the capacitance in this direction:

$$F = \frac{1}{2}\frac{\partial C(\xi)}{\partial \xi}V^2. \tag{4.13}$$

This electrostatic force can be used to drive a nanomechanical resonator. There are two main system configurations, which are discussed subsequently. The first subsection discusses the scenario where the change in capacitance in (4.13) is induced by the movement of an electrode in the nanoelectromechanial system. The second subsection focuses on the scenario where the system energy (4.11) is altered by the movement of a dielectric material inside the nanoelectromechanical system.

4.2.1.1 Forces Between Electrodes

Electrode pairs with a potential difference V feel the Coulomb force from the net charge difference. This electrostatic force can be exploited to actuate nanomechanical resonators. One way to do so is to use the mechanical resonator as one of the electrodes. That directly implies that the nanomechanical resonator has to be conductive. Either the resonator is made of a conductive material, e.g., a carbon nanotube [15, 16], graphene [17], or aluminum [18]. In case that the mechanical structure is nonconductive, such as silicon nitride, it has to be metallized, e.g., by adding a thin film of aluminum [14, 19, 20] or a sheet of graphene [14].

Depending on the geometry and arrangement of a specific electrostatic resonator design, the appropriate force model has to be used. Here three common models for specific boundary conditions are presented, as schematically depicted in Fig. 4.5.

Normal Force between Parallel Plates

The probably most commonly found electrode configuration is where a nanomechanical resonator is arranged in parallel to a surface with a potential difference V in between them. A schematic drawing is shown in Fig. 4.5a. Neglecting fringe fields at the plates' border, the capacitance between two parallel plates of area $A = wL$

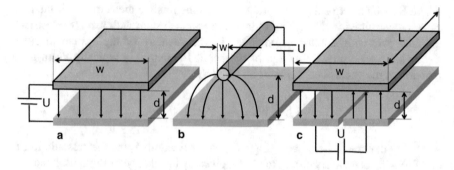

Fig. 4.5 Schematics of three electrode arrangements: (**a**) parallel plates with potential difference V, (**b**) wire and parallel surface with potential difference V, and (**c**) floating potential plate parallel to coplanar electrode pair with potential difference V

and situated in a medium with ε_r is given by

$$C = \varepsilon_0 \varepsilon_r \frac{A}{d} \tag{4.14}$$

which with (4.13) yields the force

$$F = -\frac{1}{2} \varepsilon_0 \varepsilon_r \frac{A}{d^2} V^2. \tag{4.15}$$

Normal Force between a Wire Parallel to Substrate

In this scenario, as schematically depicted in Fig. 4.5b, a thin wire with diameter w is placed in a distance $d \gg w$ from a parallel substrate (measured from the center of the wire). Such a situation is typically given for very thin resonators such as carbon nanotubes. In that case the capacitance can be approximated by

$$C = \frac{2\pi \varepsilon_0 \varepsilon_r L}{\ln\left(4\frac{d}{w}\right)} \tag{4.16}$$

which with (4.13) yields the electrostatic force normal to the wire's plane

$$F = -\frac{\pi \varepsilon_0 \varepsilon_r L}{d \ln\left(4\frac{d}{w}\right)^2} V^2. \tag{4.17}$$

In the case of a beam which neither fulfills the non-fringe assumption of the parallel plate model nor the fringe-only assumption of the wire model, the capacitance and the resulting force can be readily approximated by a combination of the two models [21]. For this purpose, the wire diameter is chosen equal to the beam's thickness $w = h$.

Normal Force between a Coplanar Electrode Pair and a Floating Electrode

In this configuration, as shown in Fig. 2.8a, a floating electrode of area A is placed in parallel over a coplanar electrode pair with a potential difference V. A schematic drawing is shown in Fig. 4.5c. An equivalent parallel plate capacitor would be half the area with double the distance d, which by neglecting fringe fields results in the capacitance [14]:

$$C = \frac{1}{4}\varepsilon_0\varepsilon_r\frac{A}{d} \qquad (4.18)$$

which with (4.13) yields the electrostatic force normal to the plate's plane

$$F = -\frac{1}{8}\varepsilon_0\varepsilon_r\frac{A}{d^2}V^2. \qquad (4.19)$$

4.2.1.2 Dielectric Polarization Force

Dielectric polarization force or *Kelvin polarization force* [22] is the force that is being exerted on a dielectric object located in a nonuniform electric field. A well-known example is the pulling of a dielectric slab into a parallel capacitor by the electric fringe field at the edge of the capacitor, an effect often discussed in standard physics books. The main advantage of the dielectric polarization force over the conventional electrostatic force between electrodes, as discussed in the previous subsection, is the possibility to work with pristine dielectric mechanical structures. The lack of metallization is interesting from many standpoints, e.g., minimal mechanical and optical losses. Figure 4.6 shows examples of applications of dielectric polarization forces for the transduction of nanomechanical resonators.

The dielectric polarization force acting on a dielectric object can readily be calculated by means of the framework introduced in the beginning of this Sect. 4.2.1 on page 112. Besides the energy approach, it is possible to derive a force term based on a dipole approximation. For this it is assumed that a dipole, which is located in an electric field $\mathbf{E}$, does not alter the field. The force acting on a single infinitesimally small dipole is then given by [24]

$$\mathbf{F} = \mathbf{p} \cdot \nabla\mathbf{E} \qquad (4.20)$$

where $\mathbf{p}$ is the dipole moment. A dielectric material can be modeled as consisting of an infinity of single noninteracting dipoles, which results in a material dependent macroscopic polarization $\mathbf{P}$. Replacing the single dipole moment with the macroscopic polarization in (4.20) then yields the *dielectric polarization force density* [22] of a dielectric material:

$$\mathbf{f}_{DPF} = \mathbf{P} \cdot \nabla \mathbf{E}. \tag{4.21}$$

With the polarization of a linear dielectric material with a susceptibility χ_e

$$\mathbf{P} = \varepsilon_0 \chi_e \mathbf{E} = \varepsilon_0 (\varepsilon_r - 1) \mathbf{E} \tag{4.22}$$

the dielectric polarization force density, after some vector algebra, becomes [22]

$$\mathbf{f}_{DPF} = \frac{1}{2} \varepsilon_0 (\varepsilon_r - 1) \nabla (\mathbf{E} \cdot \mathbf{E}). \tag{4.23}$$

Because the force density is proportional to the gradient of the electric field squared, the resulting net force always acts toward the electric field intensity maximum.

Coplanar Electrode Design

A typical resonator design based on the dielectric polarization force is based on a coplanar electrode configuration with the dielectric resonator placed on top, as can be seen in the examples in Fig. 4.6b and c. A schematic of this coplanar

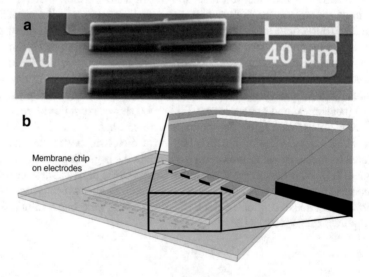

Fig. 4.6 Examples of applications of dielectric polarization forces for transduction of (**a**) micromechanical polymeric resonators (Reprint from [23], with permission from AIP Publishing), and (**b**) a nanomechanical silicon nitride resonator (schematic drawing) (Reprinted from [14], with permission from AIP Publishing)

Fig. 4.7 Schematic drawing of a coplanar electrode arrangement. The resulting nonuniform electric field exerts a dielectric polarization force on a dielectric beam in the direction of the field intensity maximum which is toward the electrode gap

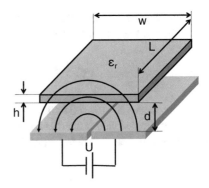

electrode configuration is schematically depicted in Fig. 4.7. For a relatively small gap between the electrodes and a homogeneous dielectric medium, the electric field lines are concentric circles, given by

$$E_0 = \frac{V}{\pi\sqrt{x^2 + y^2}}. \tag{4.24}$$

However, in the here-discussed scheme, the concentric field lines get altered by the presence of the nanomechanical dielectric beam. In order to take this field disturbance into account, the factor α is introduced. The electric field inside the dielectric beam then is given by

$$E_d = \alpha E_0 \tag{4.25}$$

where α represents the field reduction and field imperfection inside the dielectric material. With (4.23), the Kelvin polarization force density inside the dielectric beam (with ε_r) in the normal direction to the electrodes becomes:

$$
\begin{aligned}
f_{DPF} &= \frac{1}{2}\varepsilon_0(\varepsilon_r - 1)\frac{\partial E_d^2}{\partial y} \\
&= -\varepsilon_0(\varepsilon_r - 1)\frac{\alpha^2 V^2 y}{\pi^2(x^2 + y^2)^2}.
\end{aligned}
\tag{4.26}
$$

Assuming that the beam is situated in a medium such as vacuum or air with a relative dielectric constant close to unity, the force acting on the dielectric beam is readily obtained by integration of the force density over the volume of the dielectric beam:

$$
\begin{aligned}
F_{DPF} &= L\int_d^{d+h}\int_{-w/2}^{w/2} f_{DPF}\, dxdy \\
&= \frac{1}{\pi^2}\varepsilon_0(\varepsilon_r - 1)\alpha^2 V^2 L \frac{d\arctan\frac{w}{2(d+h)} - (d+h)\arctan\frac{w}{2d}}{d(d+h)}.
\end{aligned}
\tag{4.27}
$$

Assuming the beam width to be large compared to its height ($w \gg h$) and its distance to the electrodes ($w \gg d$), the force simplifies to

$$F_{DPF} \approx -\frac{1}{2\pi}\varepsilon_0(\varepsilon_r - 1)\alpha^2 L \frac{h}{d(d+h)}V^2. \tag{4.28}$$

The values for the field correction are typically in the range between $0.6 < \alpha < 1$. Numerical α values for dielectric beams with different dielectric constants ε_r and heights h can be calculated from the following heuristic formula:

$$\alpha = \frac{8301h^{0.6156}}{8301h^{0.6156} + (\varepsilon_d)^b - 1} \tag{4.29}$$

with

$$b = 0.5887 - \frac{0.003795}{h^{0.3184}}, \tag{4.30}$$

which approximates the numerical values with an average relative error of 2.0%.

When comparing the dielectric polarization force (4.28) acting on a dielectric beam to the electrostatic force acting on a floating conducting beam in vacuum (4.19)

$$F = -\frac{1}{8}\varepsilon_0 L \frac{w}{d^2}V^2 \tag{4.31}$$

it shows that the two force terms are very similar, in particular if the beam thickness is a lot smaller than the distance to the electrodes ($h \ll d$). The main difference is that the force on a dielectric beam is proportional to the beam thickness h, while the force on a conducting beam is proportional to the beam width w. For a beam with a height-to-thickness aspect ratio of unity, the forces are of similar order of magnitude. However, the width is typically larger than the thickness; this is particularly true for membrane or plate structures. In this case it is possible to design an array of coplanar electrode, which results in an interdigitated electrode design [14]. In such a design, the electrode pitch is typically of the order of the distance d. Hence, for a wide membrane or plate structure, the force on a dielectric structure is as a rough approximation a factor h/d smaller.

4.2.1.3 Electrostatic Actuation Nonlinearity

Electrostatic transduction is inherently nonlinear and is a common source of actuation nonlinearity, as mentioned in Sect. 1.3 on page 19. In that case, assuming an electrostatic force acting in normal direction of two electrodes, the driving force term that is included in (2.132) takes the form:

$$F(x, t) = -\frac{1}{2}\varepsilon_0\varepsilon_r \frac{AV(t)^2}{(d - U(x, t))^2}. \tag{4.32}$$

The force (4.32) depends on the dynamic gap between the resonator and the driving electrode, $d - U(x, t)$, and the voltage applied between them, $V(t)$. It is possible to perform a Taylor expansion of (4.32) around d:

$$F(x, t) = \frac{1}{2}\varepsilon_0\varepsilon_r \frac{AV^2(t)}{d^2}.$$

$$\cdot \left[1 + 2\frac{u(x, t)}{g_0} + 3\left(\frac{U(x, t)}{d}\right)^2 + 4\left(\frac{U(x, t)}{d}\right)^3 + \mathscr{O}\left(\frac{U(x, t)}{d}\right)\right] \tag{4.33}$$

where the nonlinear terms start to become apparent. The first term in the square bracket gives rise to the actual actuation force. The second term is a tunability term that affects directly the measured resonance frequency as discussed in Sect. 6.2.1.1, even at small amplitudes, i.e., it does not induce nonlinearity. The third and the fourth terms can be treated in a similar way to the nonlinear terms for the case of geometrical nonlinearity, yielding an equation of motion similar to (2.135), but that in this case has an additional term proportional to the displacement to the square:

$$\ddot{u}_n(t) + \frac{\Gamma_{eff,n}}{m_{eff,n}}\dot{u}_n(t) + \frac{k_{eff,n}}{m_{eff,n}}u_n(t) + \frac{\beta_{eff,n}}{m_{eff,n}}u_n^2(t) + \frac{\alpha_{eff,n}}{m_{eff,n}}u_n^3(t) = \frac{F_n(t)}{m_{eff,n}}. \tag{4.34}$$

4.2.2 Capacitively Induced Current

Besides the actuation of a nanomechanical resonator, the capacitive transduction technique can also be used to detect the motion of a vibration. The electric charge Q_e stored in the electrostatic system is given by

$$Q_e = C(\xi)V. \tag{4.35}$$

If the potential V is kept constant, a change of the capacitance causes a change of the amount of stored electric charges in the system. In other words, a mechanical vibration in ξ-direction, which modulates the capacitance, periodically "pumps" electric charges on and off the system. This charge migration can be detected as a small current. According to (4.35), this induced current can readily be increased by increasing the applied potential V, typically called the bias voltage or V_{DC}.

In a typical nanomechanical system, the mechanically induced variation of the capacitance $\Delta C \ll C_{ft}$ is a lot smaller than the total system capacitance, also

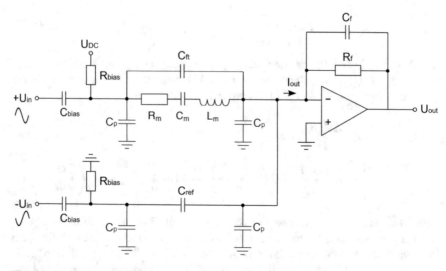

Fig. 4.8 Equivalent circuit diagram of a differential capacitive readout scheme with a transimpedance amplifier. [25]

called feedthrough capacitance C_{ft}, which remains constant. The detection of this small relative variation of the system capacitance is a main challenge of the capacitive readout scheme of the vibrational motion of a nanomechanical system. Here, different capacitive readout strategies are discussed briefly.

A common remedy is to perform a differential measurement in which the induced current from a reference capacitance, which is equal to the feedthrough capacitance ($C_{ref} = C_{ft}$), is subtracted from the signal obtained from the nanomechanical device. Like that the signal from the feedthrough capacitance disappears. A typical differential circuit is shown in Fig. 4.8. In this readout circuit, the nanomechanical resonator is substituted as an electromechanical LCR (inductor-capacitor-resistor) model.

Assuming the bias voltage V_{DC} to be much larger than the AC excitation voltage V_{in}, the equivalent impedance of the nanomechanical resonator can be described by [26]

$$Z_m(s) = R_m + i\omega L_m + \frac{1}{i\omega C_m} \tag{4.36}$$

and

$$R_m = \frac{m_{eff}\Omega}{V_{DC}^2 \left(\frac{\partial C}{\partial \xi}\right)^2 Q}$$

$$C_m = \frac{V_{DC}^2 \left(\frac{\partial C}{\partial \xi}\right)^2}{\Omega^2 m_{eff}} \tag{4.37}$$

$$L_m = \frac{m_{eff}}{V_{DC}^2 \left(\frac{\partial C}{\partial \xi}\right)^2}$$

with the eigenfrequency Ω, effective mass m_{eff}, quality factor Q, and the capacitance change per unit deflection $\partial C/\partial \xi \approx \partial C_{ft}/\partial \xi$. As depicted in the schematic in Fig. 4.8, it is an option to build a differential scheme with a passive reference capacitance C_{ref} in parallel to the nanomechanical device. Neglecting parasitic capacitances C_p, the total impedance of the nanomechanical device is the parallel combination of (4.36) and the feedthrough capacitance C_{ft} plus the reference capacitance C_{ref}, which then gives the equivalent impedance of

$$Z = \frac{V_{in}}{I_{out}} = \left[\frac{1}{Z_m} + i\omega C_{ft} - i\omega C_{ref}\right]^{-1}. \tag{4.38}$$

When measuring I_{out} directly with a device with an input impedance of $Z_f = 50\Omega$, the transfer function H of the nanoelectromechanical system becomes:

$$H = \frac{V_{out}}{V_{in}} = \frac{Z_f}{Z}. \tag{4.39}$$

An example of the transfer function (4.39) for a nanomechanical beam resonator is shown in Fig. 4.9. In the case of an unbalanced transduction scheme with $C_{ref} = 0$, the nanomechanical vibration induces a small resonance peak plus an anti-peak as a result of the interplay with the feedthrough capacitance C_{ft}. Additionally, the feedthrough capacitance produces a large background signal which overshadows the resonance peak. In the case that the feedthrough capacitance is balanced with an adjusted reference capacitance ($C_{ref} = C_{ft}$), the background signal from the feedthrough capacitance vanishes, and only the mechanically induced signal remains. From Fig. 4.9 the importance of the bias voltage V_{DC} becomes eminent. It is therefore advisable to maximize the bias voltage without destroying the nanomechanical device.

From the transfer functions plotted in Fig. 4.9, it can be seen that the obtainable signals are going to be very small. The detection of this faint capacitive current I_{out} is challenging due to impedance mismatches and parasitic capacitances C_p, which were neglected in Fig. 4.9. The root of this challenge lies in the typically ultrahigh impedance of the nanomechanical resonator (4.36). When I_{out} is measured directly with typical 50 Ω measurement equipment, the corresponding voltage

(a)   (b)

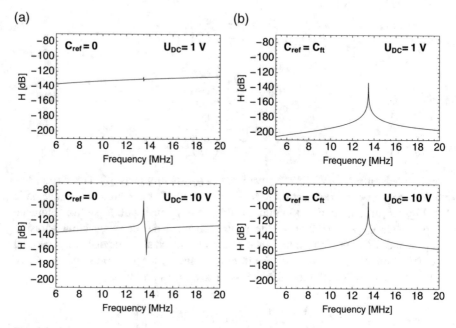

Fig. 4.9 Example of a transfer function (4.39) with unity amplification ($Z_f = 1$) of an (**a**) unbalanced and (**b**) balanced capacitive detection circuit. The measured resonance peak comes from a doubly clamped silicon nitride beam resonator from [19] with $d = 130$ nm, $L = 14\,\mu$m, $w = 200$ nm, $h = 205$ nm, $C_{ft} = 76$ aF, $\partial C/\partial \xi = 0.6$ aF/nm, $Q = 1800$, measured with $V_{DC} = 10$ V

signal gets divided by the large factor $R_m/50$ (this is the value at the resonance peaks in Fig. 4.9) and becomes very difficult to detect. When measuring with equipment with high input impedance, the small signal likely gets deteriorated by parasitic capacitances. There are several strategies on how to overcome the challenge of transducing high impedance nanomechanical resonators. Here two particular impedance mismatch strategies are quickly discussed: the use of a transimpedance amplifier and an LC filter. Similar impedance mismatch issues are also common particularly in piezoresistive (see Sect. 4.4) and piezoelectric (see Sect. 4.5) transduction schemes.

4.2.2.1 Transimpedance Amplifier

A common practice to overcome the impedance mismatch is to integrate a transimpedance amplifier in close proximity (best on chip) to the nanomechanical resonator (as shown in the schematic in Fig. 4.8) in order to save the signal from draining through the parasitic capacitances [25]. The impedance Z_f of the transimpedance amplifier is given by

$$Z_f = \frac{V_{out}}{I_{out}} = \frac{R_f}{1 + i\omega C_f R_f}. \qquad (4.40)$$

The induced current I_{out} can be amplified by choosing a large resistance R_f.

4.2.2.2 LC Filter

Another way to solve the impedance mismatch issue is to couple the nanomechanical resonator directly to an LC filter/resonator, as schematically depicted in Fig. 4.10 [19]. The equivalent impedance of this LC impedance matching circuit is given by

$$Z_{LCIM} = \left[\frac{1}{Z_m} + i\omega C_{ft} + i\omega C_{LC}\right]^{-1} + i\omega L_{LC} + R_{LC} \qquad (4.41)$$

with the impedances of the LC inductor $Z_L = sL_{LC} + R_{LC}$. The resonance frequency of the LC resonator ($\omega_{LC} = 1/\sqrt{L_{LC}C_{LC}}$) is chosen to match the frequency of the nanomechanical resonator ($\omega_{LC} = \omega_{res}$). At resonance, the total impedance seen from the measurement port is approximately given by $Z_T \approx L_{LC}/(C_{LC}R_m) + R_{LC}$[19]. The trick is now to adjust the LC components in such a way to match the impedance of the measurement equipment, which typically is 50 Ω. However, this can be hindered by the unavoidable losses R_{LC} in the inductor at room temperature. The effect of the LC filter can be seen in the reflectance plots in Fig. 4.11. Off-resonance, the LC filter reflects all energy. On-resonance however energy is absorbed and passed on to the nanomechanical resonator whose reflection peak is visible in the bottom center.

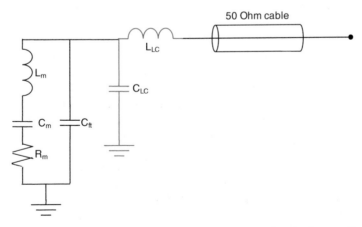

Fig. 4.10 Schematic of an LC impedance matching capacitive detection circuit, according to [19]

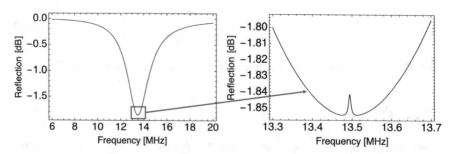

Fig. 4.11 Calculated reflection $(20 \log[(Z_{LCIM} - 50)/(Z_{LCIM} + 50)])$ of an LC impedance-matched capacitive readout circuit (4.41). The measured resonance peak comes from a doubly clamped silicon nitride beam resonator from [19] with $d = 130\,\mathrm{nm}$, $L = 14\,\mu\mathrm{m}$, $w = 200\,\mathrm{nm}$, $h = 205\,\mathrm{nm}$, $C_{ft} = 76\,\mathrm{aF}$, $\partial C/\partial \xi = 0.6\,\mathrm{aF/nm}$, $Q = 1800$, measured with $V_{DC} = 10\,\mathrm{V}$, $C_{LC} = 4.2\,\mathrm{pF}$, $L_{LC} = 33\,\mu\mathrm{H}$, $R_{LC} = 470\,\Omega$

This is a narrowband technique which requires an adjustment of the LC filter for a particular nanomechanical system. The advantage however lies in its simplicity, low temperature applicability, and possibility of multiplexing, that is, to transduce arrays of mechanical resonators via a single channel[19]. The same technique of on-resonance coupling of a nanomechanical resonator to an LC resonator has also been used inversely, that is, to use a nanomechanical resonator for signal detection [20].

4.2.3   Other Capacitive Detection Schemes

Besides the here discussed capacitive detection via the induced current, there exist other capacitive detection schemes, such as coupling of a nanomechanical resonator to a single-electron transistor [27, 28] or a microwave cavity [18, 29]. In the former technique, a nanomechanical resonator is vibrating in close proximity to the gate electrode of a transistor. The resonator is capacitively coupled to the transistor, and its vibration is detected as a modulation of its conductance. In the latter technique, the displacement of the capacitively coupled nanomechanical resonator changes the resonance frequency of a superconductive microwave cavity. In this optomechanical configuration, the mechanical resonance peak appears as a frequency modulation sideband of the cavity.

Carbon nanotube resonators are commonly transduced in semi-capacitive schemes. In these, the carbon nanotubes are spanned in close proximity over a conductive substrate which acts as a gate electrode. The vibration of the semiconductive tubes close to the gate modulates their conductivity. This transistor technique is a widespread technique for the transduction of semiconducting carbon nanotubes [16, 30–34].

4.3 Thermoelastic Actuation

A nanomechanical resonator can be actuated by local pulsed heating. The local thermal expansion of the material induces strain fields which can translate into a mechanical motion. This effect is distinct in multi-material structures with variations in the thermal expansion coefficient. In particular in bilayer beams, the heating causes a deflection due to the different thermal expansion of the different layers. In a doubly clamped beam, a pulsed heating causes a modulation of the tensile stress resulting in a parametric actuation. But even in a homogeneous material, a heating pulse can induce stress gradients which causes a net actuation force. Typically, the thermoelastic effect is maximal when heating the nanomechanical resonator at the anchoring [35]. The local heating can be induced by local photothermal heating either with a pulsed laser or with a resistive heating element, as schematically depicted in Fig. 4.12.

Optical heating is often combined with an external optical readout and allows a fully optical transduction of nanomechanical systems [35]. However, it is important to distinguish this thermoelastic technique from the optic forces used in optomechanics. The thermoelastic technique has several drawbacks, such as the net increase in temperature and the thermal relaxation time which limits the frequency bandwidth. The thermoelastic actuation with an integrated ohmic element is a technique that is particularly suitable to be combined with a piezoresistive detection (as discussed in Sect. 4.4) [36]. An example of such a combined transduction of a doubly clamped nanomechanical beam is shown in Fig. 4.13.

4.4 Piezoresistive Detection

Piezoresistive detection is based on the strain-induced change in resistance of a conductive element, typically called a *strain gauge*. For small elastic strain, the

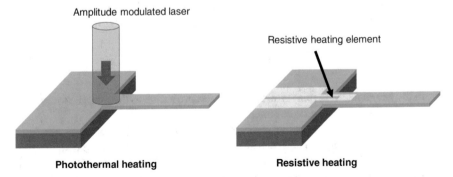

Photothermal heating **Resistive heating**

Fig. 4.12 Schematic drawing of thermoelastic actuation by local photothermal heating with an amplitude modulated laser (left) or with an integrated resistive element which can be heated by passing an electric current (right)

Fig. 4.13 (**a**) Scanning
electron micrograph of one a
single-crystal silicon carbon
resonator comprising an
80-nm-thick gold
thermoelastic bilayer actuator
(right inset) and a
30-nm-thick piezoresistive
loop of a gold palladium alloy
(left inset). (**b**) Schematic of
the transduction based on a
signal downmixing [37]. DC
denotes directional coupler,
FD frequency doubler and
HPF high-pass filter [36]

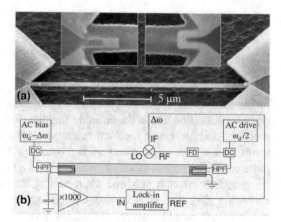

piezoresistive effect can be assumed to be linear. In this case the *gauge factor* (GF)
is defined as the relative change of the resistance R per strain ε:

$$GF = \frac{\Delta R}{R}\frac{1}{\varepsilon}. \tag{4.42}$$

The resistance of a longitudinal strain gauge of length L and a rectangular cross
section of area A is given by $R = \rho_R L/A$, with the resistivity ρ_R. The change in
resistance is given by the sum of the change of each resistance factor:

$$\Delta R = \Delta\rho_R \frac{\partial R}{\partial \rho_R} + \Delta L \frac{\partial R}{\partial L} + \Delta A \frac{\partial R}{\partial A}, \tag{4.43}$$

which results in the relative change of resistance

$$\frac{\Delta R}{R} = \frac{\Delta\rho_R}{\rho_R} + \frac{\Delta L}{L} - \frac{\Delta A}{A}. \tag{4.44}$$

Substituting the relative changes of geometry with $\Delta L/L = \varepsilon L$ and $\Delta A/A \approx
-2\nu\varepsilon$ results in the longitudinal gauge factor according to (4.42) of[38, 39]

$$GF = \frac{\Delta\rho_R}{\rho_R}\frac{1}{\varepsilon} + (1+2\nu). \tag{4.45}$$

From the gauge factor in this form, two effects contributing to the total piezoresistive
behavior in a material can be summed up.

The first term in (4.45) is the *strain coefficient of resistivity*, which comes from
the dilatation of the material. The resulting change of the interatomic distance can
alter a material's conductivity mechanism. The second term in (4.45) is a purely
geometric effect coming from the elongation and thinning of the material under
physical strain.

In bulk metals, typical values for the strain coefficient of resistivity $\frac{\Delta\rho_R}{\rho_R}\frac{1}{\varepsilon}$ range from 0.5 to 3 [40]. The geometric gauge effect typically results in values between 1.6 and 1.9. Combined, this results in typical gauge factors of bulk metals between $2 < GF < 5$. There are some exceptions, such as nickel for which large negative gauge factor have been measured ($GF = -12$)[40]. However, the gauge factor of metals is a function of sheet resistivity which can be controlled by the thickness of a metal thin film[38, 41]. It has been observed that initially GF of metal thin films decreases from the initial bulk value when thinned. A minimum of GF occurs in most metals at a sheet resistance around $10^3\,\Omega$ where the gauge factor values are significantly below the bulk values. However, above this sheet resistance, the gauge factor has shown to increase steeply for very thin films (<1 nm) with maximal values measured, e.g., for gold of $GF = 97$[38].

In semiconductors the strain coefficient of resistivity is particularly distinct, and the geometric gauge effect is secondary. Here strain alters the band gap, which changes the energy barrier an electron has to overcome to raise to the conduction band. Semiconductors, such as silicon, have gauge factors roughly ranging from $GF \approx \pm 30$ for polysilicon [42] to $GF \approx \pm 110$ for single-crystal silicon [43] depending on the doping level, doping type, temperature, and crystal orientation [39]. As in metals, in semiconductors the piezoresistivity is a function of the thickness of the strain gauge. As an example, exceptionally large gauge factors of several thousand have been observed with single-crystical silicon nanowires [44].

Piezoresistive detection is applicable for static as well as dynamic measurements. Typically, a reference and a measuring resonator are connected with two external resistors to form a Wheatstone bridge configuration [45, 46]. In this way an output signal is only recorded when there is a difference in the deflection between the two resonators. Another readout technique is based on signal downmixing [36, 37], as shown in the example in Fig. 4.13.

Instead of striving for maximizing the gauge factor, it has been shown that low-resistance gold metal strain gauges can have significant advantages over high-resistance silicon strain gauges for the transduction of nanomechanical resonators (see Fig. 4.14) [36, 47]. Even though the gauge factor of gold is relatively low compared to silicon, the final frequency resolution is high because of the low electrical noise introduced by the low-resistance gold film. Furthermore, the integrated gold electrode allows a perfect impedance matching with standard electronic equipment with resistances of 50 ohm. This allows the direct pickup of the piezoresistive signal without the need for a complex signal detection scheme.

4.5 Piezoelectric Actuation and Detection

Piezoelectricity is a material property that was first discovered by the Curie brothers circa 1880 [48]. From a physical point of view is the property of a material to generate charges when it is deformed and vice versa, upon the application of an

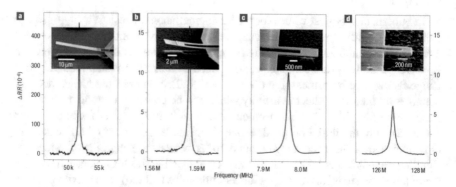

Fig. 4.14 Example of a piezoresistive detection of nanomechanical resonators with gold strain gauges. The SiC nanocantilevers were measured at room temperature in vacuum. The data plots show the fundamental-mode resonance frequencies. The inset shows SEM micrographs of the specific devices [47]

electric field, the material deforms. Mathematically, it can be seen as the coupling between the elastic and electric constitutional equations [49], which can be written down as

$$\tilde{\varepsilon}_{ij} = \tilde{S}_{ijkl}\tilde{\sigma}_{kl} + \tilde{d}_{ijk}E_k \rightarrow \varepsilon_i = S_{ij}\sigma_j + d_{ij}E_j$$
$$D_i = \in_{ij} E_j + d_{ij}^T\sigma_j \tag{4.46}$$

where ε and σ are strain and stress vectors (1×6 vectors) and their $\sim$ counterparts are tensors (3×3 matrices), D and E are the displacement and electric field, S is the compliance matrix (6×6 matrix) and $\tilde{S}$ is the fourth rank compliance tensor, $\in$ is the dielectric permittivity and d is the piezoelectric matrix (3×6 matrix) with $\tilde{d}$ the piezoelectric third rank tensor. In the absence of the latter, both equations in (4.46) remain uncoupled, which is the case for non-piezoelectric materials.

Back at the beginning of the twentieth century, much work on the piezoelectric material properties was performed, and it ended up with the invention of the piezo-electric oscillator [50], which ended up (and still is) being a fundamental part of the technological development of our society. It took many years till microfabrication processes were able to produce high-quality thin layers of piezoelectric materials [51–54]. It is this that has hampered for decades the miniaturization of piezoelectric actuation. As it can be inferred from (4.46), transduction efficiency will be strongly dependent on the magnitude of the piezoelectric tensor, thus the requirement for good materials. Many materials have been used ZnO [55], PZT [52], GaN [56, 57], GaAs[58], etc., but the preferred material due to its compatibility with standard CMOS post-processing and its overall good material properties has been aluminum nitride. The state-of-the-art of fabrication for this material shows that it can be deposited via sputtering and that layers with a good crystal alignment in the *c-axis* have been obtained even down to 10 nm-thick layers [59–61].

The governing equations for both actuation and detection are (4.46). As in any problem of structural mechanics, the actual solution of the kinematic equations, the solution for the deflection/deformation of a structure, will depend on the boundary conditions. In the case of this book, it will depend on whether the resonator/mode to be studied is flexural, torsional, bulk, 1D, 2D, etc. In each one of those cases, the resulting equations will be different. The three more used types of piezoelectric resonators are flexural beams and cantilevers [59–63], dilational (thickness) bulk acoustic wave mode [53, 56, 64], and lateral contour mode [65–67].

4.5.1 Piezoelectric Actuation

In this case one should look into the first equation in (4.46). This tells us that we need metal electrodes on top or both sides of the piezoelectric layer in order to create an electric field when applying a voltage. This electric field will modify the generalized Hooke's law [68] and forcing the structure out of the original equilibrium. For bulk mode resonators, the effect is quite direct, and one only needs to look into d_{31} and d_{33} coefficients which will directly give the deformation in the case of lateral or thickness modes, respectively (assuming that the third axis is defined out of plane and that a symmetry exists where no preferential orientation can be found within the plane). In the case of flexural modes, it is a little bit more complicated: the piezoelectric layer that is deformed via the application of voltage must not be centered around the neutral axis of the mechanical structure. Once this condition holds true, the expansion translates into a finite bending moment that, in turn, causes deflection. Equation (4.47) shows the formula for the bending moment of a cantilever (clamped-free) beam of length L and width w (see Fig. 4.15)

$$M(t) = \frac{d_{31} w z_{offset}}{S_{11,piezo}} V(t) = d_{31} E_{piezo} w z_{offset} V(t) \tag{4.47}$$

where E_{piezo} is the Young's modulus of the piezoelectric material and z_{offset} is the distance between the center of the piezoelectric layer to the neutral axis of the structure, i.e., if $z_{offset} = 0$, there is no bending moment and thus no deflection. Assuming now that the metal electrodes have a length L_{act} and a harmonic excitation at a frequency ω, it is possible to write the deflection:

$$u_n(\omega) = \frac{M(\omega) \int_0^{L_{act}} \phi_n''(x) dx}{m_{eff,n} \omega_n^2} \frac{1}{1 - \left(\frac{\omega}{\omega_n}\right)^2 + j \frac{\omega}{\omega_n Q}} \tag{4.48}$$

which we can write in terms of more basic parameters as

$$u_n(\omega) \approx \chi_n^A \frac{d_{31}E_{piezo}wz_{offset}L^2}{\langle EI \rangle} \frac{V}{1 - \left(\frac{\omega}{\omega_n}\right)^2 + j\frac{\omega}{\omega_n Q}} \tag{4.49}$$

where $\langle EI \rangle$ is the flexural rigidity of the beam, and the proportionality parameter χ_n^A needs to be calculated for each structure and mode, following

$$\chi_n^A = \frac{\int_0^{L_{act}} \phi_n''(x)\mathrm{d}x}{L^2 \int_0^L \phi_n''^2(x)\mathrm{d}x} = \frac{\phi_n'(L_{act})}{L^2 \int_0^L \phi_n''^2(x)\mathrm{d}x} \tag{4.50}$$

In the case of the first mode of a cantilever with full electrode coverage, the coefficient is $\chi_{1,cant}^A \approx 5.34$.

4.5.2 Piezoelectric Detection

In this case one should look into the second equation in (4.46), which tells us that a displacement field will be generated even in the absence of electric field or, rather, that an excess of displacement field will be generated.

As it is very well known from Maxwell's equations, if a displacement field changes over time, that induces a displacement current which for the case represented in Fig. 4.15

$$I_D(t) = \int_{A_{elec}} \frac{\partial D(t)}{\partial t}\mathrm{d}A \tag{4.51}$$

$$I_D(\omega) = j\omega \in \frac{wL_{elec}}{t_{PZE}}V_{in} + j\omega\chi_n^D d_{31}E_{piezo}\frac{wz_{offset}}{L}u_n(\omega) \tag{4.52}$$

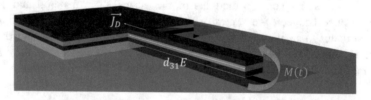

Fig. 4.15 Cartoon of a multilayer cantilever beam composed of metal and piezoelectric materials. When a voltage is applied between top and bottom electrodes, an electric field E is generated within the piezoelectric layer that is turned into expansion $d_{31}E$. When the said layer is not centered around the neutral axis of the structure, a bending moment $M(t)$ is created that causes the deflection of the cantilever. This motion creates an alternating strain and stress field that will give rise to a displacement field D. Following Maxwell's equations, an alternating displacement field also creates a displacement current $\overrightarrow{J_D}$ which will be collected

where the term χ_n^D is a proportionality term that is defined as

$$\chi_n^D = L \int_0^{L_{det}} \phi_n''(x)\,dx = L\phi_n'(L_{det}) \qquad (4.53)$$

where L_{det} is the length of the actuation electrode.

Combining (4.52) and (4.49), it is possible to reconstruct the full response of the electromechanical system, as done for the electrostatic detection in Sect. 4.2.2 on page 119

$$I_D(t) = \left(j\omega C_0 + j\omega \chi_n^A \chi_n^D \frac{(d_{31} E_{piezo} w z_{offset})^2 L}{\langle EI \rangle} \frac{1}{1 - \left(\frac{\omega}{\omega_n}\right)^2 + j\frac{\omega}{\omega_n Q}} \right) V_{in}$$

$$(4.54)$$

where we assume the actuation and detection electrodes are the same length and C_0 is the capacitance associated with the electrode. Equation (4.54) also shows the different components of the equivalent circuit, as the two terms inside the parenthesis correspond to the feedthrough capacitance and the LCR motional term in parallel. In this case it is straightforward to calculate

$$C_m = \chi_n^A \chi_n^D \frac{(d_{31} E_{piezo} w z_{offset})^2 L}{\langle EI \rangle};$$

$$L_m = \frac{1}{\omega_n^2 C_m}; \qquad (4.55)$$

$$R_m = \sqrt{\frac{L_m}{C_m}} \frac{1}{Q}.$$

Since piezoelectric detection of the motion is based on the generation of charges, an interesting option to amplify the electrical signal generated is the use of transimpedance amplifiers, as it has been described for electrostatic detection in 4.2.2.1 on page 122.

4.6 Optic Actuation and Detection

Optical transduction methods can either be fully external (off-chip) or integrated (on-chip). External techniques such as interferometry, laser-Doppler vibrometry, optical leverage readout, and radiation pressure actuation have the advantage that no physical connection to the NEMS is required with the exception of the access for the laser beam. This facilitates quick and simple testing of NEMS prototypes. In optical waveguide end coupling, evanescent light coupling, and Mach-Zehnder interferometry the optical transduction is fully integrated on-chip. Besides the coupling of an external fiber to the chip, integrated techniques need no alignment of laser beam and nanoresonator and compared to the external techniques are not limited by the diffraction of light.

4.6.1 Optical Forces

Photons bare a momentum $p = h/\lambda$, with Planck constant h and wavelength λ, which can be used to actuate nanomechanical resonators. The resulting force resulting from the momentum conservation is called *radiation pressure*, which is given by

$$p_{rad} = \frac{I_{rad}}{c} \tag{4.56}$$

where I_{rad} is the irradiance [W/m^2] and c is the speed of light. For an incident light beam of power P oriented normal to a planar surface, the resulting force is then

$$F_{rad} = \frac{P}{c}(2R + A), \tag{4.57}$$

where R and A are the reflectivity and absorptivity, respectively, of the surface. If the surface is perfect absorber ($A = 1$) and absorbs one hundred percent of the incoming radiation, then the entire photon momentum is transferred and the surface experiences the entire corresponding radiation pressure. If the surface however is a perfect mirror ($R = 1$) and reflects one hundred percent of the incident light, then the recoil from the reflected photons doubles the resulting force.

As can be readily calculated from (4.57), a reasonable laser power of $P = 1$ mW produces in a force in the pN range. This is enough to drive a high-Q micro- or nanomechanical resonator [69]. However, the total momentum can be increased to higher and more useful magnitudes by increasing the number of phonons that are reflected. This can be achieved with a high-Q optical resonator, such as a high-finesse optical cavity (the finesse describes the number of reflections a photon undergoes before it escapes the cavity). This interaction of phonons inside an optical cavity with a nanomechanical resonator is the subject of cavity optomechanics [6, 7]. One of the most important features of cavity optomechanics is the possibility to use radiation pressure to counteract the Brownian motion of a nanomechanical resonator by specifically detuning the wavelength of the phonons with respect to the high-finesse cavity. This technique allows the cooling of a nanomechanical resonator to the ground state at which it can exhibit quantum mechanical behavior [70]. Hence, cavity optomechanics is a unique tool to study quantum mechanical theory on macroscopic test objects. The techniques applied in cavity optomechanics have also successfully been applied to electrostatic systems (see Sect. 4.2.1) in so-called microwave cavity optomechanics [18]. An optical force not only acts if a light beam is reflected off of a reflective nanomechanical structure by means of momentum transfer [71] but also if a nanomechanical waveguide structure is placed in the vicinity of a nonuniform light field (*dispersive coupling*). Respective strong electromagnetic field gradients can be found in the near-field (evanescent field), e.g., of waveguides [72], optical toroidal resonators [73], or optical microsphere resonators [74].[1]

Light absorbed by the nanomechanical resonator causes a local photothermal heating. Such a heating can induce a thermoelastic response which can be used for actuation. This thermoelastic technique is discussed in Sect. 4.3 on page 125.

4.6.2 Interferometric Detection

Interferometric detection of nanomechanical motion is among the most precise detection techniques available. The principle lies in the superposition of two optical waves of the same frequency. A phase difference in one wave, induced by the interaction with a nanomechanical resonator, causes an interference pattern in the combined wave. There are several interferometric detection schemes, such as the Fabry-Pérot cavity, the Mach-Zehnder interferometer, and the Michelson interferometer, which are commonly used to detect the motion of nanomechanical resonators in a myriad of different configurations. Interferometers typically use a homodyne detection scheme, that is, the reference radiation has the same frequency as the probing beam. In contrast, laser-Doppler vibrometers typically use a heterodyne detection scheme, that is, the frequency between the reference and probing radiation differs. A fixed frequency shift is typically added to one of the beams by means of an acousto-optic modulator (Bragg cell). The four mentioned interferometric detection schemes are quickly introduced in more detail below.

4.6.2.1 Fabry-Pérot Cavity

A Fabry-Pérot cavity is the most basic interferometer consisting of one semi-transparent and one reflective plate (see schematic in Fig. 4.16a). In its most simple form, it can be formed between a suspended nanomechanical resonator and the underlying reflective substrate. The movement of the suspended structure is detected by shining a coherent light on it and detecting the change in intensity of the reflected

---

[1] There is an interesting analogy to electrostatic to be made. The optical radiation pressure observed in optical cavity transduction schemes has an electrostatic analog in the force between two electrodes or capacitor plates (see Sect. 4.2.1.1 on page 113). And the dispersive optical force experienced by a nanomechanical waveguide in a nonuniform optical field has an analog in the force acting on a dielectric nanomechanical structure that is placed in a nonuniform electric field (see Sect. 4.2.1.2 on page 115).

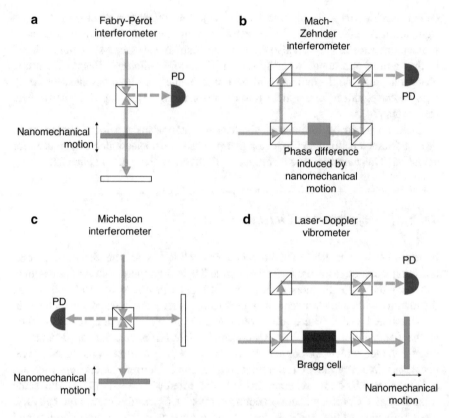

Fig. 4.16 Schematic drawings of the different types of interferometers commonly used for the detection of nanomechanical motion

light [75]. This method works well for micro-sized structures. Even though it is limited by the diffraction of light, it has been successfully used to detect the vibration of nanowires [76]. More advanced Fabry-Pérot cavities with high finesses are typically used in optomechanics [71, 77, 78].

4.6.2.2 Mach-Zehnder Interferometer

In a Mach-Zehnder interferometer (see schematic in Fig. 4.16b), the light from a single source is split into two paths which are recombined afterward. In one of the two path branches, a phase difference is induced by a nanomechanical resonator, which after recombination results in an interference [72, 73, 79].

4.6.2.3 Michelson Interferometer

In a Michelson interferometer (see schematic in Fig. 4.16c) the light from a single source (homodyne) is split and recombined after the probe light beam is reflected off a nanomechanical resonator and the reference beam is reflected off a mirror. Since a Michelson interferometer is open to thermal drifts and hence in order to constantly operate at the optimal phase, it is required to actively stabilize the phase to the mid-fringe position by means of a mirror mounted on a piezoelectric transducer in the reference arm by means of a feedback control [80].

4.6.2.4 Laser-Doppler Vibrometer

As the name implies, a laser-Doppler vibrometer detects the Doppler shift of a probing laser that is being reflected off a moving sample surface. The Doppler frequency shift of the probing wave with wavelength λ is given by

$$\Delta f_{Doppler} = 2\frac{\partial U}{\partial t}/\lambda. \tag{4.58}$$

From this fundamental effect it's obvious that a laser-Doppler vibrometer detects the velocity of a moving sample, in contrast to the other interferometric detection schemes discussed above. In order to pick up the Doppler shift, the probing beam is frequency modulated at a fixed frequency with an acousto-optic modulator. A schematic of the optical setup is depicted in Fig. 4.16d. The Doppler shift induced by the movement of the mechanical resonator is then reconstructed by demodulation of the detection signal. The advantage of a laser-Doppler vibrometer over standard interferometers is the fact that it doesn't require any phase stabilization. Laser-Doppler vibrometers are commercially available with frequency bandwidths up to 1.2 GHz with scanning capability[2]. Figure 4.17 shows a scan of a transparent silicon nitride microstring measured with an MSA-400 system from Polytec.

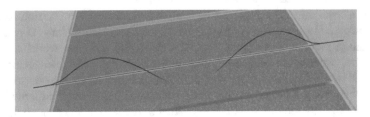

Fig. 4.17 Scan of the third bending mode of a 200-μm and 103-nm-thick silicon nitride microstring measured with MSA-500 from Polytec GmbH

---

[2] Polytec UHF-120

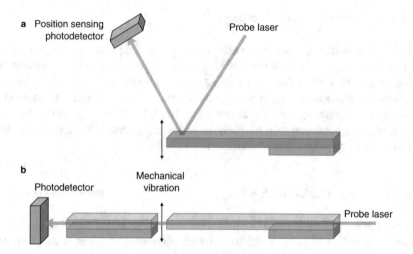

Fig. 4.18 Schematic drawings of the (**a**) optical leverage detection and the (**b**) end-coupled optical waveguide detection

4.6.3   Beam Deflection Detection

The vibrational motion of nanomechanical resonators can be detected by the deflection of a probing light beam by the resonator. Two common techniques are schematically depicted in Fig. 4.18 and briefly explained in the following paragraphs.

4.6.3.1   Optical Leverage

Optical leverage is a commonly used detection technique known from atomic force microscopy and is often used for static and dynamic detection of microcantilevers. A schematic of the technique is depicted in Fig. 4.18a. A laser is focused on the back of a cantilever. The reflected light is then detected by a position-sensitive photodetector. However, this method is rarely used for nanomechanical resonator. The method requires a resonator end-point deflection and thus works best with cantilevers, whereas in NEMS doubly clamped beams are prevalent. Another drawback is the difficult alignment of the laser such that it hits the photo-detector after having been reflected on the nanomechanical resonator. Furthermore, the resonator typically requires a reflective coating in order to reflect enough light into the position sensing photodetector.

4.6.3.2 End-Coupled Optical Waveguide

In this detection method, the nanomechanical resonator acts as a waveguide that couples at its end to another waveguide (see Fig. 4.18b). When the waveguide-cantilever is vibrating, it changes the coupling efficiency (transmission) which results in an optical amplitude modulation [81, 82]. This method enables the multiplexing of an array of resonators with a single probing laser. In order to obtain a maximal response, it is favorable if the coupled waveguides are a bit misaligned to each other. This detection method does not require a coherent light source as interferometric methods do. It is also not limited by the diffraction of light, which is a problem of the optical leverage technique.

4.6.4 Scattering Optomechanical Transduction

This technique relies on the interaction of an optical scatterer with a probing light beam. Plasmonic nanoparticles are particularly interesting scattering objects and are therefore often used. There are different implementations for the scattering optomechanical transduction. In one design, a nanomechanical vibration affects the scattering behavior, which, e.g., can be achieved with plasmonic dimers, as schematially shown in Fig. 4.19a.

Plasmonic resonances correspond to the resonant oscillation of the polarized electron cloud of metallic nanostructures. These resonance frequencies are typically located at the frequency of light in the visible to near-infrared regime. In between two such plasmonic nanostructures that are in close proximity, a strong and very localized electromagnetic field enhancement can be observed, which is often referred to as a "hotspot." These hotspots are famously used in surface-enhanced

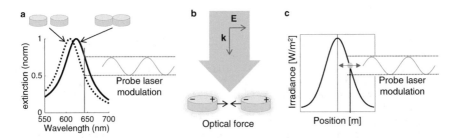

Fig. 4.19 Schematics of scattering optomechanical transduction implementations. (**a**) Readout with a plasmonic dimer antenna. An increasing gap distance causes a blue shift of the optic extinction peak, which can be detected with a probing laser. (**b**) Plasmonic forces for actuation. Localized surface plasmon resonances cause the polarization of metal nanoparticles. This creates coulomb force between the two particles of such a dimer. (**c**) Optical readout of an optical scatterer situated in a gradient optic field

Raman spectroscopy (SERS) to "focus" the probing light below the diffraction limit. It is possible to couple plasmonic resonances to the mechanical motion of nanomechanical motion. The plasmonic resonance can either be a surface plasmon resonance, supported by the surface of a gold-coated resonator [83, 84], or a localized surface plasmon resonance, supported by metallic nanoparticles [85, 86]. The plasmonic readout scheme is schematically explained in Fig. 4.19 for the case of a plasmonic dimer antenna (two metallic nanoparticles). An increase of the particle distance causes a blue shift of the plasmonic resonance peak. This shift modulates a probing laser with a wavelength located at the slope of the plasmonic resonance peak.

In addition, the plasmonic dimer implementation also allows for optomechanical actuation, as schematically depicted in Fig. 4.19b. Such electrostatic forces arise from the oscillating charge separation and the resulting coulomb attraction of negative and positive polarization centers in the plasmonic nanodimers [87]. The resulting forces are strong enough to actuate the vibration of such dimer nano-pillar resonators, which makes for a promising actuation method [88].

A second implementation of a scattering-based readout is shown in Fig. 4.19c. Here a scattering resonator is vibrating in an optical gradient field. Naturally, such a gradient field is present in any focused optical beam with a Gaussian irradiance profile. This method has been used, e.g., to detect the vibrations of carbon nanotubes [89]. In order to enhance the scattering cross section, the nanotubes feature at their tip a Pt nanoparticle deposited by focused electron beam-induced deposition. The method has also been demonstrated with laterally grown Si nanowires [90]. Such nanowires naturally comprise metal nanoparticles at their tip, which act as the catalyst during synthesis. The method can also be extended to Au nanoparticles sitting on a nanomechanical resonator, such as SiN trampolines, for the application as position-sensitive detector [91].

References

1. Scheible, D. V., Erbe, A., & Blick, R. H. (2003). Dynamic control and modal analysis of coupled nano-mechanical resonators. *Applied Physics Letters, 82*, 3333.
2. Dohn, S., Hansen, O., & Boisen, A. (2008). Measurement of the resonant frequency of nano-scale cantilevers by hard contact readout. *Microelectronic Engineering, 85*(5–6), 1390–1394.
3. Jensen, K., Weldon, J., Garcia, H., & Zettl, A. (2007). Nanotube radio. *Nano Letters, 7*(11), 3508–3511.
4. Cakmak, O., Ermek, E., Kilinc, N., Bulut, S., Baris, I., Kavakli, I. H., Yaralioglu, G. G., & Urey, H. (2015). A cartridge based sensor array platform for multiple coagulation measurements from plasma. *Lab Chip, 15*(1), 113–120.
5. Suter, M., Ergeneman, O., Zürcher, J., Schmid, S., Camenzind, A., Nelson, B. J., & Hierold, C. (2011). Superparamagnetic photocurable nanocomposite for the fabrication of microcantilevers. *Journal of Micromechanics and Microengineering, 21*(2), 025023.
6. Kippenberg, T. J., & Vahala, K. J. (2008). Cavity optomechanics: Back-action at the mesoscale. *Science, 321*(5893), 1172–6.

7. Aspelmeyer, M., Kippenberg, T. J., & Marquard, F. (2014). Cavity optomechanics. *Reviews of Modern Physics, 86*(4), 1391–1452.

8. Cleland, A. N., & Roukes, M. L. (1996) Fabrication of high frequency nanometer scale mechanical resonators from bulk Si crystals. *Applied Physics Letters, 69*, 2653.

9. Venstra, W. J., Westra, H. J. R., Gavan, K. B., & der Zant, H. S. J. (2009). Magnetomotive drive and detection of clamped-clamped mechanical resonators in water. *Applied Physics Letters, 95*(26), 263103.

10. Feng, X. L., White, C. J., Hajimiri, A., & Roukes, M. L. (2008). A self-sustaining ultrahigh-frequency nanoelectromechanical oscillator. *Nature Nanotechnology, 3*(6), 342–346.

11. Cleland, A. N., & Roukes, M. L. (1999). External control of dissipation in a nanometer-scale radiofrequency mechanical resonator. *Sensors and Actuators A: Physical, 72*(3), 256–261.

12. Schwab, K. (2002). Spring constant and damping constant tuning of nanomechanical resonators using a single-electron transistor. *Applied Physics Letters, 80*(7), 1276–1278.

13. Tang, H. X., Huang, X. M. H., Roukes, M. L., Bichler, M., & Wegscheider, W. (2002). Two-dimensional electron-gas actuation and transduction for GaAs nanoelectromechanical systems. *Applied Physics Letters, 81*(20), 3879–3881.

14. Schmid, S., Bagci, T., Zeuthen, E., Taylor, J. M., Herring, P. K., Cassidy, M. C., Marcus, C. M., Guillermo Villanueva, L., Amato, B., Boisen, A., Cheol Shin, Y., Kong, J., Sørensen, A. S., Usami, K., & Polzik, E. S. (2014). Single-layer graphene on silicon nitride micromembrane resonators. *Journal of Applied Physics, 115*(5), 054513.

15. Eichler, A., Chaste, J., Moser, J., & Bachtold, A. (2011). Parametric amplification and self-oscillation in a nanotube mechanical resonator. *Nano Letters, 11*(7), 2699–2703. PMID: 21615135.

16. Lassagne, B., Garcia-Sanchez, D., Aguasca, A., & Bachtold, A. (2008). Ultrasensitive mass sensing with a nanotube electromechanical resonator. *Nano Letters, 8*(11), 3735–3738.

17. Weber, P., Güttinger, J., Tsioutsios, I., Chang, D. E., & Bachtold, A. (2014). Coupling graphene mechanical resonators to superconducting microwave cavities. *Nano Letters, 14*(5), 2854–2860.

18. Teufel, J. D., Li, D., Allman, M. S., Cicak, K., Sirois, A. J., Whittaker, J. D., & Simmonds, R. W. (2011). Circuit cavity electromechanics in the strong-coupling regime. *Nature, 471*(7337), 204–208.

19. Truitt, P. A., Hertzberg, J. B., Huang, C. C., Ekinci, K. L., & Schwab, K. C. (2007). Efficient and sensitive capacitive readout of nanomechanical resonator arrays. *Nano Letters, 7*(1), 120–126.

20. Bagci, T., Simonsen, A., Schmid, S., Villanueva, L. G., Zeuthen, E., Appel, J., Taylor, J. M., Sørensen, A., Usami, K., Schliesser, A., & Polzik, E. S. (2014). Optical detection of radio waves through a nanomechanical transducer. *Nature, 507*(7490), 81–85.

21. Yuan, C. P., & Trick, T. N. (1982). A simple formula for the estimation of the capacitance of two-dimensional interconnects in VLSI circuits. *IEEE Electron Device Letters, 3*(12), 391–393.

22. Haus, H. A., & Melcher, J. R. (1989). *Electromagnetic fields and energy.* Englewood Cliffs, NJ: Prentice Hall.

23. Schmid, S., Wendlandt, M., Junker, D., & Hierold, C. (2006). Nonconductive polymer microresonators actuated by the Kelvin polarization force. *Applied Physics Letters, 89*(16), 163506.

24. Jones, T. B. (1995). *Electromechanics of {p}articles.* Cambridge: Cambridge University Press.

25. Schmid, S. (2009). *Electrostatically actuated all-polymer microbeam resonators—Characterization and Application, volume 6 of Scientific Reports on Micro and Nanosystems.* Allemagne: Der Andere Verlag.

26. Nguyen, C. T. -C., & Howe, R. T. (1999). An integrated {CMOS} micromechanical resonator high-{Q} oscillator. *IEEE Journal of Solid-State Circuits, 34*(4), 440–455.

27. Knobel, R. G., & Cleland, A. N. (2003). Nanometre-scale displacement sensing using a single electron transistor. *Nature, 424*(6946), 291–293.

28. LaHaye, M. D., Buu, O., Camarota, B., & Schwab, K. C. (2004). Approaching the quantum limit of a nanomechanical resonator. *Science, 304*(5667), 74.
29. Regal, C. A., Teufel, J. D., & Lehnert, K. W. (2008). Measuring nanomechanical motion with a microwave cavity interferometer. *Nature Physics, 4*(7), 555–560.
30. Witkamp, B., Poot, M., & Van Der Zant, H. S. J. (2006). Bending-mode vibration of a suspended nanotube resonator. *Nano Letters, 6*(12), 2904–2908.
31. Sazonova, V., Yaish, Y., Üstünel, H., Roundy, D., Arias, T. A., & McEuen, P. L. (2004). A tunable carbon nanotube electromechanical oscillator. *Nature, 431*(7006), 284–287.
32. Gouttenoire, V., Barois, T., Perisanu, S., Leclercq, J. L., Purcell, S. T., Vincent, P., & Ayari, A. (2010). Digital and FM demodulation of a doubly clamped single-walled carbon-nanotube oscillator: Towards a nanotube cell phone. *Small, 6*(9), 1060–1065.
33. Chaste, J., Eichler, A., Moser, J., Ceballos, G., Rurali, R., & Bachtold, A. (2012). A nanomechanical mass sensor with yoctogram resolution. *Nature Nanotechnology, 7*(5), 301–304.
34. Moser, J., Güttinger, J., Eichler, A., Esplandiu, M. J., Liu, D. E., Dykman, M. I., & Bachtold, A. (2013). Ultrasensitive force detection with a nanotube mechanical resonator. *Nature Nanotechnology, 8*(7), 493–496.
35. Ilic, B., Krylov, S., Aubin, K., Reichenbach, R., & Craighead, H. G. (2005). Optical excitation of nanoelectromechanical oscillators. *Applied Physics Letters, 86*, 193114.
36. Bargatin, I., Kozinsky, I., & Roukes, M. L. (2007). Efficient electrothermal actuation of multiple modes of high-frequency nanoelectromechanical resonators. *Applied Physics Letters, 90*(9), 093116.
37. Bargatin, I., Myers, E. B., Arlett, J., Gudlewski, B., & Roukes, M. L. (2005). Sensitive detection of nanomechanical motion using piezoresistive signal downmixing. *Applied Physics Letters, 86*(13), 1–3.
38. Parker, R. L., & Krinsky, A. (1963). Electrical resistance-strain characteristics of thin evaporated metal films. *Journal of Applied Physics, 34*(9), 2700–2708.
39. Kanda, Y. (1982). A graphical representation of the piezoresistance coefficients in silicon. *IEEE Transactions on Electron Devices, 29*(1), 64–70.
40. Kuczynski, G. C. (1954). Effect of elastic strain on the electrical resistance of metals. *Physical Review, 94*(1), 61–64.
41. Jen, S. U., Yu, C. C., Liu, C. H., & Lee, G. Y. (2003). Piezoresistance and electrical resistivity of Pd, Au, and Cu films. *Thin Solid Films, 434*(1-2), 316–322.
42. French, P. J., & Evans, A. G. R. (1989) Piezoresistance in polysilicon and its applications to strain gauges. *Solid State Electronics, 32*(1), 1–10.
43. Smith, C. S. (1954). Piezoresistance effect in germanium and silicon. *Physical Review, 94*(1), 42–49.
44. He, R., & Yang, P. (2006). Giant piezoresistance effect in silicon nanowires. *Nature Nanotechnology, 1*(1), 42–46.
45. Yu, X., Thaysen, J., Hansen, O., & Boisen, A. (2002). Optimization of sensitivity and noise in piezoresistive cantilevers. *Journal of Applied Physics, 92*(10), 6296–6301.
46. Boisen, A., Thaysen, J., Jensenius, H., & Hansen, O. (2000). Environmental sensors based on micromachined cantilevers with integrated read-out. *Ultramicroscopy, 82*(1-4), 11–16.
47. Li, M., Tang, H. X., & Roukes, M. L. (2007). Ultra-sensitive NEMS-based cantilevers for sensing, scanned probe and very high-frequency applications. *Nature Nanotechnology, 2*(2), 114–120.
48. Curie, J., & Curie, P. (1880). Development by pressure of polar electricity in hemihedral crystals with inclined faces. *Bulletin de la Societe de Mineralogique de France, 3*, 90.
49. Cady, W. G. (1964). *Piezoelectricity: An introduction to the theory and applications of electromechanical phenomena in crystals* (new rev. ed.). New York: Dover Publications.
50. Cady, W. G. (1922). The piezo-electric resonator. *Proceedings of the Institute of Radio Engineers, 10*(2), 83–114.

51. Racine, G. A., Muralt, P., & Dubois, M. A. (1998). Flexural-standing-wave elastic force motor using ZNO and PZT thin film on micromachined silicon membranes for wristwatch applications. *Smart Materials & Structures, 7*(3), 404–416.
52. Dubois, M. A., & Muralt, P. (1998). PZT thin film actuated elastic fin micromotor. *IEEE Transactions on Ultrasonics Ferroelectrics and Frequency Control, 45*(5), 1169–1177.
53. Dubois, M. A., Muralt, P., & Sagalowicz, L. (1999). Aluminum nitride thin films for high frequency applications. *Ferroelectrics, 224*(1-4), 671–678.
54. Martin, F., Muralt, P., Dubois, M. A., & Pezous, A. (2004). Thickness dependence of the properties of highly *c-axis* textured ALN thin films. *Journal of Vacuum Science & Technology A, 22*(2), 361–365.
55. Itoh, T., & Suga, T. (1993) Development of a force sensor for atomic force microscopy using piezoelectric thin films. *Nanotechnology, 4*, 218.
56. Ansari, A., & Rais-Zadeh, M. (2014). A thickness-mode AlGaN/GaN resonant body high electron mobility transistor. *IEEE Transactions on Electron Devices, 61*(4), 1006–1013.
57. Ansari, A., Liu, C. Y., Lin, C. C., Kuo, H. C., Ku, P. C., & Rais-Zadeh, M. (2015). GaN micromechanical resonators with meshed metal bottom electrode. *Materials, 8*(3), 1204–1212.
58. Masmanidis, S. C., Karabalin, R. B., De Vlaminck, I., Borghs, G., Freeman, M. R., & Roukes, M. L. (2007). Multifunctional nanomechanical systems via tunably coupled piezoelectric actuation. *Science, 317*(5839), 780–783.
59. Karabalin, R. B., Matheny, M. H., Feng, X. L., Defay, E., Le Rhun, G., Marcoux, C., Hentz, S., Andreucci, P., & Roukes, M. L. (2009). Piezoelectric nanoelectromechanical resonators based on aluminum nitride thin films. *Applied Physics Letters, 95*(10), 103111.
60. Sinha, N., Wabiszewski, G. E., Mahameed, R., Felmetsger, V. V., Tanner, S. M., Carpick, R. W., & Piazza, G. (2009). Piezoelectric aluminum nitride nanoelectromechanical actuators. *Applied Physics Letters*, 95(5), 053106.
61. Zaghloul, U., & Piazza, G. (2014). Synthesis and characterization of 10 nm thick piezoelectric ALN films with high *c-axis* orientation for miniaturized nanoelectromechanical devices. *Applied Physics Letters, 104*(25), 253101.
62. Ivaldi, P., Abergel, J., Matheny, M. H., Villanueva, L. G., Karabalin, R. B., Roukes, M. L., Andreucci, P., Hentz, S., & Defay, E. (2011). 50 nm thick ALN film-based piezoelectric cantilevers for gravimetric detection. *Journal of Micromechanics and Microengineering, 21*(8), 085023.
63. Karabalin, R. B., Villanueva, L. G., Matheny, M. H., Sader, J. E., & Roukes, M. L. (2012). Stress-induced variations in the stiffness of micro- and nanocantilever beams. *Physical Review Letters*, 108(23), 236101.
64. O'Connell, A. D., Hofheinz, M., Ansmann, M., Bialczak, R. C., Lenander, M., Lucero, E., Neeley, M., Sank, D., Wang, H., Weides, M., Others, O'Connell, A. D., Wenner, J., Martinis, J. M., & Cleland, A. N. (2010). Quantum ground state and single-phonon control of a mechanical resonator. *Nature, 464*(7289), 697–703.
65. Piazza, G., Stephanou, P. J., & Pisano, A. P. (2006). Piezoelectric aluminum nitride vibrating contour-mode mems resonators. *Journal of Microelectromechanical Systems, 15*(6), 1406–1418.
66. Piazza, G., Stephanou, P. J., & Pisano, A. P. (2007). Single-chip multiple-frequency ALN MEMS filters based on contour-mode piezoelectric resonators. *Journal of Microelectromechanical Systems, 16*(2), 319–328.
67. Piazza, G., Stephanou, P. J., & Pisano, A. P. (2007). One and two port piezoelectric higher order contour-mode MEMS resonators for mechanical signal processing. *Solid-State Electronics, 51*(11-12), 1596–1608.
68. Gere, J. M., & Goodno, B. J. (2013). *Mechanics of materials* (8th ed.). Stamford, CT: Cengage Learning.
69. Ma, D., Garrett, J. L., & Munday, J. N. (2015). Quantitative measurement of radiation pressure on a microcantilever in ambient environment. *Applied Physics Letters, 106*(9), 091107.

70. Chan, J., Alegre, T P M., Safavi-Naeini, A. H., Hill, J. T., Krause, A., Gröblacher, S., Aspelmeyer, M., & Painter, O. (2011). Laser cooling of a nanomechanical oscillator into its quantum ground state. *Nature, 478*(7367), 89–92.
71. Gröblacher, S., Hammerer, K., Vanner, M. R., & Aspelmeyer, M. (2009). Observation of strong coupling between a micromechanical resonator and an optical cavity field. *Nature, 460*(7256), 724–727.
72. Li, M., Pernice, W. H. P., Xiong, C., Baehr-Jones, T., Hochberg, M., & Tang, H. X. (2008) Harnessing optical forces in integrated photonic circuits. *Nature, 456*(7221), 480–484.
73. Anetsberger, G., Arcizet, O., Gavartin, E., Unterreithmeier, Q. P., Weig, E. M., Kotthaus, J. P., & Kippenberg, T. J. (2010). Near-field cavity optomechanics with nanomechanical oscillators. In: *Lasers and Electro-Optics (CLEO) and Quantum Electronics and Laser Science Conference (QELS), 2010 Conference on* (vol. 5(12), pp. 1–9).
74. Brawley, G. A., Vanner, M. R., Larsen, P. E., Schmid, S., Boisen, A., & Bowen, W. P. (2016). Nonlinear optomechanical measurement of mechanical motion. *Nature Communications, 7*(1), 1–7.
75. Carr, D. W., Sekaric, L., & Craighead, H. G. (1998). Measurement of nanomechanical resonant structures in single-crystal silicon. *Journal of Vacuum Science & Technology B: Microelectronics and Nanometer Structures, 16*(6), 3821–3824.
76. Belov, M., Quitoriano, N. J., Sharma, S., Hiebert, W. K., Kamins, T. I., & Evoy, S. (2008). Mechanical resonance of clamped silicon nanowires measured by optical interferometry. *Journal of Applied Physics, 103*, 74304.
77. Thompson, J. D., Zwickl, B. M., Jayich, A. M., Marquardt, F., Girvin, S. M., & Harris, J. G. E. (2008). Strong dispersive coupling of a high-finesse cavity to a micromechanical membrane. *Nature, 452*(7183), 72–75.
78. Andrews, R. W., Peterson, R. W., Purdy, T. P., Cicak, K., Simmonds, R. W., Regal, C. A., & Lehnert, K. W. (2014). Bidirectional and efficient conversion between microwave and optical light. *Nature Physics, 10*(April), 321–326.
79. Verhagen, E., Deléglise, S., Weis, S., Schliesser, A., & Kippenberg, T. J. (2012). Quantum-coherent coupling of a mechanical oscillator to an optical cavity mode. *Nature, 482*(7383), 63–67.
80. Barg, A., Tsaturyan, Y., Belhage, E., Nielsen, W. H. P., Møller, C. B., & Schliesser, A. (2017). Measuring and imaging nanomechanical motion with laser light. *Applied Physics B, 123*(1), 8.
81. Li, M., Pernice, W. H. P., & Tang, H. X. (2009). Broadband all-photonic transduction of nanocantilevers. *Nature Nanotechnology, 4*(6), 377–382.
82. Nordström, M., Zauner, D. A., Calleja, M., Hübner, J., & Boisen, A. (2007). Integrated optical readout for miniaturization of cantilever-based sensor system. *Applied Physics Letters, 91*, 103512.
83. Thijssen, R., Verhagen, E., Kippenberg, T. J., & Polman, A. (2013). Plasmon nanomechanical coupling for nanoscale transduction. *Nano Letters, 13*, 3293–3297.
84. Ou, J. -Y., Plum, E., Zhang, J., & Zheludev, N. I. (2013). An electromechanically reconfigurable plasmonic metamaterial operating in the near-infrared. *Nature Nanotechnology, 8*(4), 252–255.
85. Thijssen, R., Kippenberg, T. J., Polman, A., & Verhagen, E. (2015). Plasmomechanical resonators based on dimer nanoantennas. *Nano Letters, 15*(6), 150507141241008.
86. Song, J. -H., Raza, S., van de Groep, J., Kang, J. -H., Li, Q., Kik, P. G., & Brongersma, M. L. (2021). Nanoelectromechanical modulation of a strongly-coupled plasmonic dimer. *Nature Communications, 12*(1), 1–7.
87. Raziman, T. V., Wolke, R. J., & Martin, O. J. F. (2015). Optical forces in nanoplasmonic systems: How do they work, what can they be useful for? *Faraday Discussions, 178*, 421–434.
88. Buch, Z. & Schmid, S. (2022). Design considerations of gold nanoantenna dimers for plasmomechanical transduction. *Optics Express, 30*(4), 5294–5303.

89. Tavernarakis, A., Stavrinadis, A., Nowak, A., Tsioutsios, I., Bachtold, A., & Verlot, P. (2018). Optomechanics with a hybrid carbon nanotube resonator. *Nature Communications, 9*(1), 1–8.

90. Molina, J., Ramos, D., Gil-Santos, E., Escobar, J. E., Ruz, J. J., Tamayo, J., Paulo, Á. S., & Calleja, M. (2020). Optical transduction for vertical nanowire resonators. *Nano Letters, 20*(4), 2359–2369.

91. Chien, M. -H., Steurer, J., Sadeghi, P., Cazier, N., & Schmid, S. (2020). Nanoelectromechanical position-sensitive detector with picometer resolution. *ACS Photonics, 7*(8), 2197–2203.

Chapter 5
Measurements and Noise

The potential impact of nanomechanical resonators in the field of sensing has been delineated in the previous chapter. However, the outstanding responsivity that is associated with their small sizes is not enough to make a good sensor. Indeed, it is typically the sensitivity that contains information about the sensor performance, as it determines the minimum detectable amount of the target magnitude. As a consequence, it is of the utmost importance to analyze the different noise sources that enter these systems so that a full picture can be formed upon their usability in actual applications. In this chapter, different noise sources are analyzed, starting by a study of noise affecting the motion of the device (thermomechanical noise) or the conversion of this motion into an electrically readable signal (transduction-related noise). Following the transduction chain, amplifier noise is then studied. This is shown schematically in Fig. 5.1. Finally, the implications of these noise sources into the determination of the device's frequency are taken into consideration, as well as other noise sources directly affecting the frequency. In this chapter, *power spectral density* (PSD) S is used to quantify the noise and describes the mean noise per Hz of bandwidth. Only *single-sided* noise densities are used (this is the magnitude spectrum analyzers measure).

5.1 Amplitude Noise

This section describes the different noise sources that affect the determination of the amplitude of motion, both in magnitude and phase. It can also be referred to as "*open loop*" as a contrast to the next section where the effect of the different noise sources in the determination of frequency is going to be described.

© The Author(s), under exclusive license to Springer Nature Switzerland AG 2023
S. Schmid et al., *Fundamentals of Nanomechanical Resonators*,
https://doi.org/10.1007/978-3-031-29628-4_5

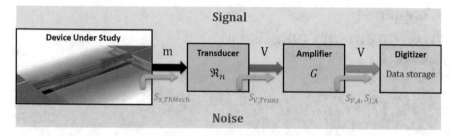

Fig. 5.1 Schematic showing the different noise sources along the different transduction stages. Within the resonator itself, still in the mechanical domain thermomechanical noise enters the system. Within the transduction to electrical signal, different noises may enter the system, e.g., Johnson-Nyquist, Hooge, and shot noise. Finally the amplifying stage of the system will insert the input noise of the amplifier(s), dominated by current and voltage noise of the first amplifier

5.1.1   Fundamentals

Before starting the description of the different types of noise that can enter into NEMS resonators, it is interesting to first take a look into two basic and fundamental concepts: transduction chain noise transfer and *noise referred to input* (RTI).

5.1.1.1   Transduction Chain Noise Transfer

It can be deduced from Fig. 5.1 that the final noise in the measurement (digital data) will have contributions from the different noises in the transduction chain. But such contributions have different influence on the final noise. Using the notation in the figure, one can write that the total noise of the signal arriving at the digitizer is

$$S_{V,total} = G_V^2 \left(S_{V,A} + S_{V,Trans} + \mathfrak{R}_n^2 \left(S_{x,ThMech} \right) \right), \tag{5.1}$$

with the mode-specific transducer responsivity $\mathfrak{R}_n$, amplifier voltage gain G_V, thermomechanical noise $S_{x,ThMech}$, transduction noise $S_{V,Trans}$, and amplifier noise $S_{V,A}$. Equation (5.1) can be generalized to any transduction chain. For example, if a second amplifier was to be used, the final noise would be given by

$$S_{V,total} = G_2^2 \left(S_{V,A2} + G_1^2 \left(S_{V,A1} + S_{V,Trans} + \mathfrak{R}_n^2 \left(S_{x,ThMech} \right) \right) \right) \tag{5.2}$$

where it becomes clear that the noise in the first amplifier is the most important one within the amplifying chain.

5.1.1.2 Noise Referred to Input (RTI)

A direct consequence from the equations in the previous paragraph is that the final noise $S_{V,total}$ depends directly on the transduction chain. This is the reason why, in order to compare transduction chains from a noise perspective, the magnitude that is taken to perform such comparison is the RTI. By comparing two different noise RTIs, one can see which transduction chain "adds" more noise to the signal.

The concept of noise RTI is easy to understand and calculate. It is first important to substitute the transduction gain (with multiple stages) by a single stage with an *equivalent* gain and an *equivalent* noise at the input of the said single stage. Like that, the system shown in Fig. 5.1 has an equivalent gain of: $G_{equiv} = G_1 \mathfrak{R}_n$ and a noise RTI given by

$$S_{RTI} = \frac{S_{V,total}}{G_{equiv}^2} = \frac{S_{V,A} + S_{V,Trans}}{\mathfrak{R}_n^2} + S_{x,ThMech}. \tag{5.3}$$

Or in the case that two amplifiers were to be used, the noise RTI becomes

$$S_{RTI} = \frac{S_{V,total}}{G_2^2 G_1^2 \mathfrak{R}_n^2} = \frac{S_{V,A2}}{G_1^2 \mathfrak{R}_n^2} + \frac{S_{V,A1} + S_{V,Trans}}{\mathfrak{R}_n^2} + S_{x,ThMech}. \tag{5.4}$$

As it can be seen in both (5.3) and (5.4), the contribution of each stage in the amplifying chain will be divided by the overall gain just until that stage, which makes the stages closest to the sensor/device the most important ones in terms of noise.

5.1.2 Thermomechanical Fluctuations

Any system that has a dissipation is subjected to damping, as is shown in Chap. 3. In the particular case of mechanical resonators, this dissipation can be seen as the coupling (interaction) between a cavity and a thermal reservoir full of randomly distributed phonons. This coupling, due to the randomness of the phonons in the reservoir, causes a noisy force that is applied to the mechanical resonator itself, as it is explained by the *fluctuation-dissipation theorem* (FDT) [1, 2]. Here, the thermomechanical fluctuation of a nanomechanical resonator is derived by means of an effective lumped-element model [3].

The equipartition theorem implies that in thermal equilibrium, a mechanical system of one degree of freedom has an average total energy (the sum of kinetic and elastic energy) in each mode that is equal to $k_B T$, where k_B is the Boltzmann constant and T is the temperature. Assuming a lumped-element model and consider the elastic energy, this means

$$\frac{1}{2}k_B T = \frac{1}{2}k_{eff}\langle z_{th}^2\rangle = \frac{1}{2}m_{eff}\omega_0^2\langle z_{th}^2\rangle \tag{5.5}$$

where $\langle z_{th}^2\rangle$ is the average thermomechanical displacement, which becomes

$$\langle z_{th}^2\rangle = \frac{k_B T}{m_{eff}\omega_0^2}. \tag{5.6}$$

Since the mean of the displacement is zero $\langle z_{th}\rangle = 0$, the average thermomechanical power can be treated as the variance of the displacement $\sigma_z^2 = \langle z_{th}^2\rangle$.

When performing noise measurements, the magnitude that is typically measured is the *power spectral density* (PSD). The average power (variance) of a signal can be calculated by integrating the corresponding PSD $S_z(\omega)$ with units [m²/Hz] over the respective frequency range. For a one-sided PSD when the frequency range is undefined the integral goes from zero to infinity:

$$\sigma_z^2 = \langle z_{th}^2\rangle = \frac{1}{2\pi}\int_0^\infty S_z(\omega)d\omega = \int_0^\infty S_z(2\pi f)df. \tag{5.7}$$

The factor $1/(2\pi)$ is because the integrating variable is an angular frequency $\omega = 2\pi f$. It falls away if the integral is calculated for a frequency f.

The force noise acting on the nanomechanical resonator is random as a function of time. As a result, the force noise is white and has a power spectral density S_F that is independent of frequency. The force noise is acting on the one degree of freedom of the nanomechanical resonator. The resulting thermomechanical noise is then given by

$$S_z(\omega) = |\chi(\omega)|^2 S_F \tag{5.8}$$

with the resonator's force susceptibility $\chi(\omega)$, which approximates to a Lorentzian function (1.34 on page 10)

$$|\chi(\omega)|^2 \approx \frac{1}{4m_{eff}^2\omega_0^2}\left[\frac{1}{(\omega_0 - \omega)^2 + \left(\frac{\omega_0}{2Q}\right)^2}\right]. \tag{5.9}$$

As a result, the spectral noise density itself is non-white due to the frequency dependency of the resonance response of a nanomechanical mode. Inserting (5.6) and (5.8) into (5.7) and integrating yields a value for the white force noise spectral density with units [N²/Hz]:

$$S_F = \frac{4m_{eff}\omega_0 k_B T}{Q}. \tag{5.10}$$

From (5.8) it is now also possible to extract the noise magnitude at the peak of the Lorentzian, which is at $\omega = \omega_0$:

$$S_{z_{th}} = S_z(\omega_0) = \frac{4k_B T Q}{m_{eff}\omega_0^3}. \tag{5.11}$$

It is worth noting that, even though the maximum value of $S_{z_{th}}$ is linearly proportional to the quality factor of the resonator Q, the overall strength of the thermal white noise S_F is inversely proportional to Q. In other words, for a given resonator, high Q means less noise overall but a higher (and narrower) noise peak.

An example of a measured thermomechanical noise peak is shown in Fig. 5.2. The peak value is given by the white noise spectral density $S_{z_{th}}$ (5.11). The transduction-based noise floor $S_{z_{tr}} = \mathcal{K}^2 S_{z_{th}}$ can be described with respect to the thermomechanical noise spectral density with the proportionality factor $\mathcal{K}$ [4]. Transduction-related noise is discussed in Sect. 5.1.3.

5.1.2.1 Amplitude Calibration and Thermometry

Before continuing to the next section, it is also interesting to mention the utility of thermomechanical noise to calibrate the responsivity of the transduction mechanism selected to detect the motion of the resonator. No matter the transduction mechanism, the magnitude that is actually measured in the lab is electrical power coming

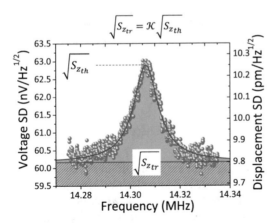

Fig. 5.2 Example of a thermomechanical noise measurement (scattered data). The fitting to a Lorentzian function (purple line) with the maximum value $\sqrt{S_{z_{th}}}$ (5.11) and a background (red line) with a value $\sqrt{S_{z_{tr}}} = \mathcal{K}\sqrt{S_{z_{th}}}$ provide information about the different noise sources in the measurement: system noise (including transduction and amplifier noise) for the flat background and thermomechanical noise for the peak. Even though the measurement is performed in volts (left axis), using the procedure described above, it is possible to calibrate the transduction back into meters (right axis)

into the analyzer (e.g., spectrum, network, or lock-in amplifier). Thus, displacement (in meters) is transduced into an electrical signal (volts), and it is necessary to find an experimental way to estimate this transduction coefficient, which can be called responsivity and in general is different for each resonator mode:

$$\Re_n = \frac{\partial V_n}{\partial u_n}. \tag{5.12}$$

One of the typical ways of estimating $\Re_n$ is to use thermomechanical noise, and the fact that the frequency dependence (5.8) and the overall integral along the frequency spectrum (5.7) of such noise are both known. The procedure starts by measuring the noise of the system under study around the resonance frequency with a spectrum analyzer. The result should always be in units of noise power (e.g., $\frac{V^2}{Hz}$) so that the size of the resolution bandwidth does not modify the final result. Once this measurement has been performed, it is necessary to fit the data to a Lorentzian function (1.34) with a background that accounts for other noise sources that are analyzed later in this chapter. This background can typically be considered as white noise (flat in the frequency domain), assuming that the measurement is narrow range around the resonant frequency of the device. The results for the fitting parameters can be equated to the expected value(s) from (5.8), and like that one can obtain the value for the transduction responsivity, $\Re_n$ (Fig. 5.2).

Once the readout amplitude is calibrated, e.g., for a measurement at room temperature, the thermomechanical PSD (5.8) can be used to estimate the temperature of a specific resonant mode. Again, this is best done by fitting the measured noise PSD with (5.8), which readily allows the extraction of T, which represents the effective temperature of the specific mode that is being analyzed. This method is particularly useful in situations where, e.g., a specific mode is actively cooled either by means of a negative feedback or laser cooling, and hence the mode temperature differs from the bath temperature.

5.1.3 Transduction-Related Noise

As it is shown in Chap. 4, there are many different ways to detect the motion of a mechanical device and transduce it into electrical energy. This of course affects the value of the first responsivity $\Re_n$ that is shown in Fig. 5.1, but it also determines the transduction noise, $S_{V_{tr}}$ or $S_{I_{tr}}$, which is the noise in the electrical domain that is inserted in the system due to the transduction technique being used. This noise S_{tr} might have itself several possible origins. Johnson-Nyquist thermal noise, Hooge $1/f$ noise, and shot noise are generally the main contributions which will be analyzed in the following paragraphs.

5.1.3.1 Johnson-Nyquist Thermal Noise

Electrical thermal noise is caused by the random thermally excited vibration of the charge carriers in a conductor. This is, in some way, similar to the Brownian motion of particles or the previously described thermomechanical motion of the resonator.

Thermal noise was first observed by J.B. Johnson and contemporaneously explained by H. Nyquist [2]. It can be thus generalized via the fluctuation-dissipation theorem (FDT) [1, 2], because a conductor has a dissipative part, i.e., the real part of the impedance or $Re(Z)$. The noise is white up to a high frequency limit, where a roll-off occurs, allowing for the *overall* noise level to remain finite. The noise magnitude, already estimated by Johnson [5], in units of power spectral density is

$$S_{V_{th}} = 4k_B T R \quad \left[\frac{V^2}{Hz}\right]$$

$$S_{I_{th}} = \frac{4k_B T}{R} \quad \left[\frac{A^2}{Hz}\right]$$

(5.13)

depending whether the noise is modeled as a current source in parallel to a noiseless resistance or a voltage source in series with a noiseless resistance. In (5.13), k_B is the Boltzmann constant, T is the temperature of the resistor in Kelvin and R is the value of the resistance or, in the more general case, the value for the real part of the impedance. In reality, as said above, the noise is not completely white as (5.13) shown, but presents a cutoff frequency beyond which the noise relaxes down, as shown in Fig. 5.3. This cutoff is fundamentally determined via the thermal occupancy of states or by the fastest time scale on which electrons are typically scattered. A more *mundane* reason for the roll-off is the parasitic capacitance that is always present in parallel to a resistor, thus limiting the amount of noise that is transferred to other parts of the circuit.

Fig. 5.3 Scaled thermal noise power spectral density versus scaled frequency. The noise is white, i.e., constant, all the way until the cutoff frequency where it drops down to zero

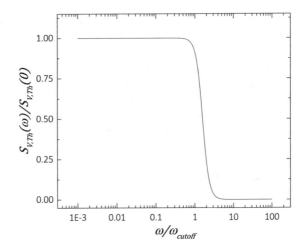

As for any type of noise, in order to calculate the actual voltage or current noise in the measurement, one needs to integrate the power spectral density over the bandwidth, which will be defined in particular for the said experiment. The bandwidth of a measurement establishes which part of the frequency spectrum is allowed to enter the detector. A very large bandwidth will thus mean a larger noise but, at the same time, a shorter integration time. On the other hand, having a small bandwidth will imply much smaller noise but a larger integration time, which can lead (but not necessarily) to a slower response time.

The overall noise formula is given by

$$V_{th} = \sqrt{\int_{f_0}^{f_1} 4k_B T R \, df} = \sqrt{4k_B T R \, BW}$$

$$I_{th} = \sqrt{\int_{f_0}^{f_1} \frac{4k_B T}{R} \, df} = \sqrt{\frac{4k_B T}{R} \, BW}$$

(5.14)

where $BW = f_1 - f_0$ is the bandwidth of the measurement and it is assumed that the cutoff frequency described in Fig. 5.3 is much larger than f_1, which is typically the case.

To finalize this section, it is necessary to highlight that this analysis presented here is *strictly* only valid when the resistor is not biased. In other words, it only works when the electrons are, in average, not moving. In the biased case, there can be a departure from the predicted values of (5.14); see [6] for more details. In addition, as soon as the resistors are biased, two more noises enter to play: shot noise and Hooge ($1/f$) noise.

5.1.3.2 Shot Noise

Current flowing is not smooth nor continuous. In the end, due to the fact that charge is quantized, one can consider current as the succession of discontinuous pulses, one for each arriving carrier. A useful analogy can be the raindrops of a heavy rain falling on a tin roof.

Shot noise is the most fundamental model of discontinuous noise in physical systems. It is not restricted to electrical systems but can be also found in any system with quantized carriers like optical systems (photons), communication systems (packets), acoustic systems (phonons), etc.

Shot noise was first described by Campbell [7, 8], followed by many studies [9], out of which the experiments done by Schottky in 1918 represent an important milestone [10]. It is associated with partitioned systems, like when current flows across a potential barrier, e.g., in diodes, transistors, vacuum tubes, etc. The spectrum is white and the intensity is given by

$$S_{I_{shot}} = 2e\zeta\, I_{bias} \quad \left[\frac{A^2}{Hz}\right]$$

$$S_{V_{shot}} = 2e\zeta\, I_{bias}\, R^2 \quad \left[\frac{V^2}{Hz}\right],$$

(5.15)

where e is the unitary electron charge, I_{bias} is the current that passes through the noisy part of resistance R and ζ is the *Fano* or correlation factor, which is a proportionality factor that ranges from 0 when the material is a perfect conductor to 1 for systems with full partition, e.g., for tunnel junctions, semiconductor fully depleted regions, etc. For the rest of cases, $0 < \zeta < 1$. In the particular case of disordered conductors, it is possible to prove that $\zeta \to \frac{1}{3}$ [11].

From (5.15) it is possible to calculate the actual value of the noise by following the same procedure shown in (5.14):

$$I_{shot} = \sqrt{\int_{f_0}^{f_1} 2e\zeta\, I_{bias}\, df} = \sqrt{2e\zeta\, I_{bias}\, BW}$$

$$V_{shot} = \sqrt{\int_{f_0}^{f_1} 2e\zeta\, I_{bias}\, R^2\, df} = \sqrt{2e\zeta\, I_{bias}\, R^2\, BW},$$

(5.16)

where once again BW is the bandwidth of the measurement. It is interesting to note the dependence of the noise intensity with the square root of the current itself, meaning that the larger the current, the larger the noise (remember the analogy with the heavy rain on the rooftop, where more rain would imply more noise).

5.1.3.3 Hooge ($1/f$) "Flicker" Noise

Up to now, the three noises that have been analyzed in this chapter have been white (thermomechanical noise is white in force; Johnson and shot noises are white in current). On the other hand, $1/f$ noise has a power spectral density that increases, in principle without limit, as frequency decreases. The first observations of this noise were done on vacuum tubes, and they caused flickering in the plate current, hence the name *flicker* noise.

This noise has several unique properties, and it is quite ubiquitous, not only in electronic systems (in particular in disordered conductors), but more generally in most physical systems, there is always an increase in the noise at low frequencies. Even though it generally is called $1/f$, this is an abuse of language. As it has been widely reported [12, 13], the reality is that the noise has a power dependence of the type $1/f^\alpha$ where $\alpha = 1$ is the most common value. In electronic circuits, the exponent is generally found within a range, typically $0.8 < \alpha < 1.3$. But there are many other systems that show mechanisms with larger values of α, e.g., fluctuations of the Earth's rotation frequency have $\alpha = 2$, and galactic radiation noise has $\alpha =$

2.7. From now on, for simplicity, the most common value for the exponent will be taken.

The origin of this type of noise in electronic components seems to stem from fluctuations of the conductance/resistance of the devices. Hooge [14] built a semi-empirical law to model this noise, which is given by

$$S_{V_{1/f}} = \frac{\gamma}{N_c} \frac{V_{bias}^2}{f} \quad \left[\frac{V^2}{Hz}\right]$$

$$S_{I_{1/f}} = \frac{\gamma}{N_c} \frac{V_{bias}^2}{R^2 f} \quad \left[\frac{A^2}{Hz}\right],$$

(5.17)

where γ is a proportionality constant to be empirically estimated for a given device, N_c is the number of carriers in the device and V_{bias} is the bias voltage applied to the noisy part of resistance R. Equation (5.17) hints the important temperature dependence that this noise has, as the number of carriers (N_c) depends strongly on temperature, plus the proportionality constant has also been observed to depend on temperature. This, in turn, makes this type of noise extremely dependent on the device and operating conditions that are being used for an experiment.

To estimate the actual noise contribution of this type, it is necessary to integrate the power spectrum density over the bandwidth

$$V_{1/f} = \sqrt{\int_{f_0}^{f_1} \frac{\gamma}{N_c} \frac{V_{bias}^2}{f} \, df} = V_{bias} \sqrt{\frac{\gamma}{N_c} \ln\left(\frac{f_1}{f_0}\right)}$$

$$I_{1/f} = \sqrt{\int_{f_0}^{f_1} \frac{\gamma}{N_c} \frac{V_{bias}^2}{R^2 f} \, df} = \frac{V_{bias}}{R} \sqrt{\frac{\gamma}{N_c} \ln\left(\frac{f_1}{f_0}\right)}.$$

(5.18)

A possible theoretical derivation for this noise can be found in the literature [15] and is based on the superposition of multiple relaxation processes, each of them with a Lorentzian-like power spectrum but different relaxation rates. After averaging, the characteristic power law $1/f$ is recovered.

5.1.3.4   Noise Equivalent Circuit

Before moving on to the amplifier noise, it is interesting to show how the noise sources that have just been analyzed look like when a circuit needs to be considered, how to place them so that one can perform a proper noise analysis and translate such noise to other places in the circuit. The two basic schematics that can be used are either the Thevenin and the Norton equivalent circuit, as it is shown in Fig. 5.4.

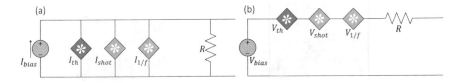

Fig. 5.4 Noise equivalent circuits for a biased resistive element. Either option, Norton (**a**), or Thevenin (**b**) can be used to calculate the overall noise and to estimate how much this noise affects the next stage in the system

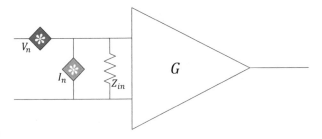

Fig. 5.5 Noise equivalent circuit to describe the ensemble of effects that are present internally in any amplifier. V_n and I_n are respectively the voltage and current noise of the amplifier, G is its gain, and Z_{in} is its input impedance which in general can take complex values. The amplifier in the model maintains the same gain as the original one, but in this case it is an *ideal* amplifier, which means that is noiseless and with an infinite input resistance

5.1.3.5 Amplifier Noise

As illustrated in Fig. 5.1, after the transduction of the motion into an electrical signal, an amplifying stage is typically used to bring the signal to a level that can be detected or digitized by other equipments in the lab. Amplifiers are very complex devices that contain many components, which makes it virtually impossible to perform a thorough and detailed analysis of how each of these components affects the overall noise within the amplifier. A noise model is helpful to simplify noise analysis at the circuit/system level, i.e., how much the *noise referred to input* is. The most extended noise model for amplifiers, which is also the simplest that allows a full analysis, is the one that includes voltage and current noise in the input of a noiseless amplifier (see Fig. 5.5 for a schematic of the equivalent circuit).

Once the schematic for the amplifier noise model is set, it is possible to calculate the noise referred to input for this amplifier when other parts are connected to it. As a simple example, let's consider the circuit presented in Fig. 5.6, where an input signal V_{in} is sent to an amplifier through a resistive line of value R, which has an associated noise $V_{n,R}$. The actual gain of the whole (total) system, which can be defined as $G_t = \frac{V_{out}}{V_{in}}$, is different than the gain of the ideal amplifier present in the schematic. It is rather straightforward to obtain that

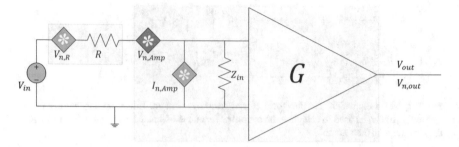

Fig. 5.6 Noise equivalent circuit for an amplifier connected to a source and a resistive element. The amplifier model is the one present in Fig. 5.5

$$V_{out} = G \frac{Z_{in}}{R + Z_{in}} V_{in} \rightarrow G_t = G \frac{Z_{in}}{R + Z_{in}}. \qquad (5.19)$$

In order to estimate the noise referred to input, it is necessary to consider independently every noise source present in the schematic of Fig. 5.6. To do this, one needs to "null" (short-circuiting voltage sources and leaving current sources open) all sources in the system except for one and calculate the noise contribution and add these contributions squared. Following this approach it is easy to obtain the noise in the output of the amplifier

$$V^2_{n,out} = \left(V^2_{n,R} + V^2_{n,A} \right) G^2 \left(\frac{Z_{in}}{R + Z_{in}} \right)^2 + I^2_{n,A} G^2 \left(\frac{R \, Z_{in}}{R + Z_{in}} \right)^2. \qquad (5.20)$$

Combining (5.20) and (5.19), one can obtain the noise referred to input of the whole system

$$V_{n,RTI} = \sqrt{ \left(V^2_{n,R} + V^2_{n,A} \right) + I^2_{n,A} R^2 }. \qquad (5.21)$$

It needs to be pointed out here that all the previous equations are strictly valid only if there is no correlation between voltage and current noise. In reality, it is possible to find some correlation between the two of them, which complicates the analysis requiring an extra term with the cross product of voltage and current noise [16].

Following with the system schematized in Fig. 5.6, one can now post the question about how does the signal to noise ratio evolve from before and after the amplifier. The signal to noise ratio (SNR) is a dimensionless figure of merit that allows to estimate the quality (cleanliness) of your measurement technique, being better, the larger the SNR. The power SNR before and after the amplifier is, respectively, given by

$$\left(\frac{S}{N}\right)_{in} = \left(\frac{V_{in}}{V_{n,R}}\right)^2 \tag{5.22}$$

and

$$\left(\frac{S}{N}\right)_{out} = \left(\frac{V_{out}}{V_{n,out}}\right)^2$$

$$= \frac{G^2 \left(\frac{Z_{in}}{R+Z_{in}}\right)^2 V_{in}^2}{\left(V_{n,R}^2 + V_{n,A}^2\right) G^2 \left(\frac{Z_{in}}{R+Z_{in}}\right)^2 + I_{n,A}^2 G^2 \left(\frac{R\,Z_{in}}{R+Z_{in}}\right)^2}$$

$$= \frac{V_{in}^2}{V_{n,R}^2 + V_{n,A}^2 + I_{n,A}^2 R^2}$$

$$= \frac{V_{in}^2}{V_{n,RTI}^2}. \tag{5.23}$$

The ratio between these two SNR defines the *noise factor*, F of the amplifier, which is given by

$$F = \frac{\left(\frac{S}{N}\right)_{in}}{\left(\frac{S}{N}\right)_{out}} = \frac{V_{n,R}^2 + V_{n,A}^2 + I_{n,A}^2 R^2}{V_{n,R}^2} = 1 + \frac{V_{n,A}^2 + I_{n,A}^2 R^2}{V_{n,R}^2} \tag{5.24}$$

where $F \geq 1$, and that it is equal to 1 when the amplifier does not add any noise to the measurement, the *ideal* case. One could argue that this implies that the amplifier only makes the measurement *worse*, as the SNR is smaller after the amplifier than before. But when considering the final stage of the measurement chain, e.g., a digitizer or an analog acquisition device, there are parts which will be noisier than the noise coming from the transduction (in the example $V_{n,R}$) and also noisier than the amplifier. Thus, the selection of a good amplifier for the first step in the amplifying stage is critical.

5.1.3.6 Noise Figure and Noise Temperature

As it has been shown in previous paragraphs, the noise factor determines how good or bad an amplifier is. There are two other parameters that are often used to describe this point: noise figure (NF) and noise temperature (T_{Amp}) which are defined as follows

$$NF = 10 \ \log_{10}(F) \tag{5.25}$$

and

$$T_{Amp} = T_0 \left(F - 1 \right), \tag{5.26}$$

where T_0 is typically taken to be 290 K. The noise figure is a positive magnitude that in the case of the ideal amplifier tends to zero, and is expressed in dB. Noise temperature is a useful magnitude that uses the concept of noise being proportional to temperature (in power), as it has been shown in the case of thermomechanical and Johnson noise. It gives an idea of how much "hotter" the system is after the amplifier.

An important point to address is that with neither of these parameters, noise factor (F), noise figure (NF), and noise temperature (T_{Amp}), it is not possible to have a direct estimation of the amplifier noise, but only a ratio to $V_{n,R}$. However, the latter noise is typically considered to be Johnson noise of a 50 Ω resistor at 290 K (unless specified otherwise), which is the most common impedance value for transmission lines. Still, this does not allow to distinguish between current and voltage noise in the amplifier, but only gives a concrete value for $V_{n,A}^2 + I_{n,A}^2 R^2$. It should also be noted that (5.24) and subsequent expressions are written as a function of noise, not power spectral density. However, one can consider the noise contributions from the amplifier as white within the measurement bandwidth. Thus, (5.24) becomes

$$F = 1 + \frac{S_{V,A} + S_{I,A} R^2}{S_R}. \tag{5.27}$$

5.2   Noise in Resonance Frequency Measurements

In the previous section of the chapter, the noise in the determination of the amplitude of motion has been analyzed. Recalling from Chap. 6, many of the applications that use a NEMS resonator as sensor involve the measurement of the resonance frequency rather than the amplitude. Hence, it is very important to study how to determine the noise in the frequency measurement.

It's clear that the somewhat extended version of the frequency detection that involves the acquisition of successive amplitude vs. frequency curves, as shown in Fig. 5.7a, is not the typical (nor optimal) way of performing this detection. In this case, the time between two consecutive experimental data points, i.e., two consecutive determinations of a device's natural frequency, takes very long, as many points need to be acquired to be able to fit each curve and extract the frequency. Therefore, open-loop frequency tracking schemes (OLS) are typically performed as schematically depicted in Fig. 5.7b. In this scheme, the slightly damped nanomechanical resonator is driven with an external signal generator at a fixed frequency $\omega \approx \omega_0$ close to its resonance frequency. This situation is depicted in Fig. 5.8. A shift of the resonance frequency then produces a shift of the corresponding phase φ that is detected, e.g., with a lock-in amplifier. Based on the known phase slope (1.24 on page 8), the corresponding resonance frequency

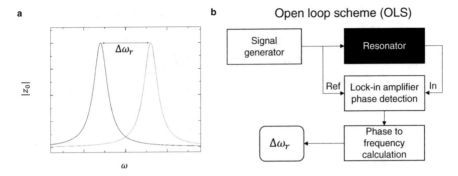

Fig. 5.7 (**a**) Frequency shift determination by subsequent frequency scanning. (**b**) Typical open-loop frequency-shift detection scheme

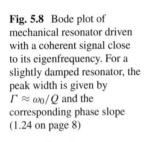

Fig. 5.8 Bode plot of mechanical resonator driven with a coherent signal close to its eigenfrequency. For a slightly damped resonator, the peak width is given by $\Gamma \approx \omega_0/Q$ and the corresponding phase slope (1.24 on page 8)

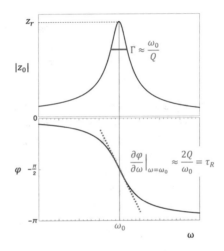

change $\Delta\omega_r$ can then readily be calculated. This method is easy to implement but comes with two major drawbacks. First, the dynamic range is limited by the line-width Γ of the resonance. Specifically, the frequency shifts have to be smaller than Γ. Second, the driving frequency is fixed and defined by the signal generator. In this situation, the response time of the system is determined by the time constant of the resonator itself $\tau_R = 2Q/\omega_0$, as discussed in Sect. 1.1.3.2 on page 10. In particular for high-Q resonators, this results in a slow response to changes in both external force and even resonance frequency [3, 17], which can be limiting for many applications.

The limitations of the open-loop scheme can be overcome with closed-loop schemes. In a closed loop system, the nanomechanical resonator acts as the frequency determining element. Hence, a closed-loop system has a larger dynamic range (it can follow frequency changes larger than the resonator linewidth), and the response time for parameter-induced frequency changes is not limited by the

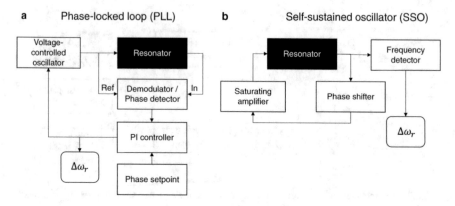

Fig. 5.9 Closed-loop frequency-shift detection schemes

resonator time constant. Changes of the resonator parameters (6.2 on page 176), such as mass and stiffness, change the resonance frequency instantaneously. The response time is limited by the respective closed-loop bandwidth. The two most common closed-loop schemes are *phase-locked loops* (PLL) and *self-sustained oscillators* (SSO) as schematically shown in Fig. 5.9. All open- and closed-loop schemes are discussed in more detail in the subsequent subsections.

The noise in a resonance frequency measurement is characterized by the Allan variance, which is introduced below in Sect. 5.2.1. In an experiment, the Allan variance can be calculated from the measured steady-state frequency noise. It is also possible to estimate the frequency noise of a NEMS resonator. Frequency noise is created by thermomechanical (Sect. 5.1.2 on page 147) and transduction noise (Sect. 5.1.3 on page 150) and can be calculated for a specific frequency tracking scheme, as will be discussed in Sect. 5.2.2 on page 162. There are additional frequency noise sources, of which thermal fluctuation noise is the most significant noise source that can affect resonator performance, which will be discussed in the last Sect. 5.2.3 on page 172.

5.2.1 Allan Variance

Here, the magnitude is introduced that is typically used as a metric for the frequency noise of a resonant sensors. This is the so-called Allan variance, named after its inventor David W. Allan [18, 19]. In principle this can be applied to any magnitude as it is defined very generally

$$\sigma_y^2(\tau) = \frac{1}{2}\left\langle \left(\bar{y}_{i+1,\tau} - \bar{y}_{i,\tau}\right)^2\right\rangle = \frac{1}{2(N-1)} \sum_{i=1}^{N-1} \left(y_{i+1,\tau} - y_{i,\tau}\right)^2, \qquad (5.28)$$

where $\bar{y}_i$ is the ith measurement of the magnitude y which is taken during an average time τ, i.e.,

$$\bar{y}_{i,\tau} = \frac{1}{\tau} \int_{(i-1)\tau}^{i\tau} y(t)dt. \tag{5.29}$$

The final result as a function of the integration time τ provides a magnitude that, in the presence of white noise, reduces when τ increases. Indeed this shows the major advantage of this variance over the classical variance. The Allan variance converges for most of the commonly encountered kinds of noise, whereas the classical variance does not always converge to a finite value. As an example, $1/f$ noise has a classical variance which does not converge, and it is commonly occurring in nature and, in particular, clock-like systems (like oscillators or PLLs).

Equations (5.28) and (5.29) are very general as they can be applied to any dimensionless variable y. In the particular case of resonance-based sensors, y is chosen to be

$$y(t) = \frac{\omega(t) - \omega_n}{\omega_n}, \tag{5.30}$$

which is the normalized instantaneous frequency difference to the natural frequency of the resonator.

The Allan variance is the best known of the time-domain magnitudes to characterize frequency inaccuracy. To know more about the other variances, the reader is directed to follow the Appendixes in Ref. [20] and the references therein. It is important to address the fact that in addition to these time-domain magnitudes, there exist their frequency-domain counterparts, which can be defined as

$$\sigma_y^2(\tau) = \frac{1}{2\pi} \frac{8}{\tau^2} \int_0^\infty \frac{[\sin(\frac{\omega\tau}{2})]^4}{\omega^2} S_y(\omega)d\omega, \tag{5.31}$$

where $S_y(\omega)$ is the fractional frequency noise spectral density with units [1/Hz]

$$S_y(\omega) = S_{y_{thm}}(\omega) + S_{y_{thf}}(\omega) + \dots \tag{5.32}$$

where $S_{y_{thm}}(\omega)$ and $S_{y_{thf}}(\omega)$ are the fractional frequency noise spectral densities due to thermomechanical/transduction noise and thermal fluctuations, respectively, which will be discussed in the subsequent Sects. 5.2.2 on the following page and 5.2.3 on page 172. There exist other frequency noise sources that potentially could contribute to the total frequency noise, such as adsorption-desorption noise [21] or photothermal back action noise [17], which have to be considered for a specific NEMS.

Depending on the application domain, time-domain or frequency-domain magnitudes are preferred. In the sensing community, it is rather the Allan variance, in the time domain, that is used so that one can directly estimate the noise in the system for a given integration time.

The spectral density of the relative frequency fluctuations $S_y(\omega)$ can be measured experimentally. Knowing $S_y(\omega)$, it is possible to calculate the expected Allan deviation with (5.31). Allan deviation (σ_y) for most common frequency tracking systems and specific parameters is calculated and discussed in the following subsection. However, the actual limitation for the detection sometimes can't be estimated that easily. Therefore, in order to actually determine a system's limits, the Allan deviation must be measured experimentally. Once that is done, one can easily locate which is the most effective point for operation. This might coincide with the minimum, but it can also happen that the application requires faster response time (smaller integration time τ), and thus resolution is compromised in favor of a faster measurement.

In the case of white frequency noise with a constant one-sided power spectral density S_y, the Allan variance (5.31) simply reduces to

$$\sigma_y^2(\tau) = \frac{S_y}{2\tau} \quad \text{(for } S_y = \text{constant)}. \tag{5.33}$$

The resulting Allan deviation has a slope $\sigma_y(\tau) \propto 1/\sqrt{\tau}$, which is characteristic for white frequency noise.

5.2.2   Frequency Noise from Thermomechanical and Transduction Noise

Assuming a steady-state situation in which the resonance frequency remains constant, open-loop and closed-loop schemes are fundamentally similar. In all schemes, the nanomechanical resonator is driven with a coherent driving signal close to its resonance frequency, as depicted in Fig. 5.8. This allows for a common discussion of the frequency noise in all these systems. The following frequency noise discussion is highly simplified, and it assumes that the resonator operates in its linear regime. More detailed and intricate discussions can be found in these selected papers [3, 4, 21, 22].

After the coherent drive of the nanomechanical resonator at resonance, there are two main noise sources that have to be considered, thermomechanical noise $S_{z_{th}}$ and transduction noise $S_{z_{tr}}$, as discussed in the previous Sect. 5.1 on page 145. In an optimal measurement, the thermomechanical noise is resolved and larger than transduction noise. Assuming white detection noise, it can be defined as a fraction of the white thermomechanical noise [4]:

$$S_{z_{tr}} = \mathcal{H}^2 S_{z_{th}}. \tag{5.34}$$

Fig. 5.10 Amplitude to phase noise transformation visualized in the phase space with the in-phase and out-of-phase quadratures Z_1 and Z_2, respectively. The amplitude and phase noise are given by their respective variances σ_z^2 and σ_φ^2

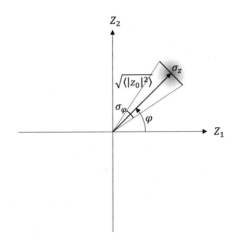

The dimensionless factor $\mathscr{K}$ corresponds to the ratio of the measured (undriven) thermomechanical noise $\sqrt{S_{z_{th}}}$ to the background/transduction noise. Such an experimental thermomechanical noise peak is shown in Fig. 5.2 where the purple and red line correspond to the thermomechanical and background noise, respectively. If $\mathscr{K} < 1$, the thermomechanical noise is resolved, and if $\mathscr{K} > 1$, transduction noise is dominating.

The combined amplitude noise (thermomechanical noise plus transduction noise) results in phase noise of the coherent resonator response. The amplitude to phase noise translation is best studied in phase space, as shown in Fig. 5.10. It shows the coherent resonator vibration at resonance with a root mean square (rms) amplitude $\sqrt{\langle |z_0(\omega_0)|^2\rangle} = z_r/\sqrt{2}$ and phase φ. The noise-induced amplitude deviation manifests in the phase plane as a sphere at the tip of the coherent drive vector. The radius of this noise sphere is given by the square root of the average noise power, which can be expressed by the variance σ_z^2 (5.7). From Fig. 5.10 it can be seen that the phase deviation is given by

$$\sigma_\varphi = \frac{\sqrt{2}\sigma_z}{z_r}, \tag{5.35}$$

which presents the translation of amplitude noise into phase noise. Hence, the noise power spectral density of thermomechanical and transduction noise (5.34) is given by

$$S_{\varphi_{th}} = \frac{2S_{z_{th}}}{z_r^2}$$

$$S_{\varphi_{tr}} = \frac{2S_{z_{tr}}}{z_r^2} = \mathscr{K}^2 S_{\varphi_{th}}. \tag{5.36}$$

The total phase noise is then the sum of the thermomechanical and transduction noise weighted by individual steady-state transfer functions $H_{\varphi_{th}}(i\omega)$ & $H_{\varphi_{tr}}(i\omega)$ that are specific to the frequency tracking scheme:

$$S_\varphi(\omega) = |H_{\varphi_{th}}(i\omega)|^2 S_{\varphi_{th}} + |H_{\varphi_{tr}}(i\omega)|^2 S_{\varphi_{tr}}$$

$$= \left(|H_{\varphi_{th}}(i\omega)|^2 + \mathscr{K}^2 |H_{\varphi_{tr}}(i\omega)|^2 \right) S_{\varphi_{th}}. \tag{5.37}$$

$H_{\varphi_{th}}(i\omega)$ & $H_{\varphi_{tr}}(i\omega)$ will be discussed for each frequency tracking scheme in the subsequent subsections.

The phase noise (5.37) directly converts into a frequency noise via the phase slope relation (1.24 on page 8), as shown in Fig. 5.8:

$$S_{\Delta\omega}(\omega) = \frac{\omega_0^2}{4Q^2} S_\varphi(\omega) = \frac{1}{\tau_R^2} S_\varphi(\omega), \tag{5.38}$$

with τ_R being the resonator time constant (1.1.3.2 on page 10). In the open-loop scheme (OLS), this conversion is done as a calculation (see Fig. 5.7b). In the closed-loop schemes (PLL & SSO), this conversion is performed by the resonator itself as depicted in Fig. 5.8.

Finally, with (5.38), (5.37), and (5.36), the fractional frequency ($y = \Delta\omega/\omega_0$) noise spectral density can be written as

$$S_{y_{thm}}(\omega) = \frac{S_{\Delta\omega}(\omega)}{\omega_0^2} = \frac{1}{2Q^2} \frac{S_{z_{th}}}{z_r^2} \left[|H_{\varphi_{th}}(i\omega)|^2 + \mathscr{K}^2 |H_{\varphi_{tr}}(i\omega)|^2 \right]. \tag{5.39}$$

As it will be shown in the discussion of each frequency tracking scheme, the transfer functions (5.37) all have low-pass characteristics with $H_{\varphi_{th}}(0) = H_{\varphi_{tr}}(0) = 1$ [4]. Hence, the relative frequency noise spectral density (5.39) for a steady state and not limited by any system bandwidth is then given by

$$S_{y_{thm}}(0) = \frac{1}{2Q^2} \frac{S_{z_{th}}}{z_r^2} \left[1 + \mathscr{K}^2 \right], \tag{5.40}$$

which with (5.11) writes as

$$S_{y_{thm}}(0) = \frac{2k_B T}{m_{eff} Q \omega_0^3 z_r^2} \left[1 + \mathscr{K}^2 \right]. \tag{5.41}$$

This equation reveals that relative frequency fluctuations can be minimized either by lowering the resonator's temperature T or by maximizing the vibrational amplitude z_r to enhance the signal-to-noise ratio. The maximization of the vibrational amplitude is limited by the onset of nonlinearity, which is given by the critical amplitude z_c (1.63 on page 23). Increasing the vibrational amplitude beyond z_c results in a bifurcation of the resonator response. After bifurcation, multiple stable points arise, there is no longer a one-to-one correspondence between the phase and amplitude, and also there is a conversion of amplitude noise into phase noise (essentially making the noise nonspherical). Assuming a thermomechanically well-resolved transduction ($\mathcal{K} \ll 1$) and setting $z_r = z_c$, (5.41) yields the critical relative frequency noise spectral density of

$$S_{y_{thm_c}}(0) \approx \frac{3\sqrt{3}}{4} \frac{k_B T \alpha_{eff}}{m_{eff}^2 \omega_0^5}, \tag{5.42}$$

where α_{eff} is the effective nonlinear stiffness of the resonator. It has been shown that this limit can be surpassed when in the nonlinear regime by moving the operation point away from a phase $\varphi = -\pi/2$ to an optimal operation point [23, 24]. Contrary to intuition, $S_{y_{thm_c}}$ is independent of Q. That is, if the measurement noise is limited by thermomechanical noise and the nanomechanical resonator is driven to its critical amplitude at the onset of nonlinearity, then the optimal relative frequency noise becomes independent of Q.

As $S_{y_{thm}}(0)$ (5.40) is white frequency noise with a constant power spectral density, the Allan variance can readily be calculated with (5.33)

$$\sigma_y^2(\tau) = \frac{S_{y_{thm}}(0)}{2\tau} = \frac{1}{4Q^2} \frac{S_{z_{th}}}{z_r^2} \frac{1}{\tau} \left[1 + \mathcal{K}^2\right]. \tag{5.43}$$

It is important to note that this expression for the Allan variance is only valid for averaging times τ that are larger than the time constant of the respective frequency tracking system, which all have a low-pass characteristic [4]. If τ is shorter than the system time constant, the white noise gets filtered by the system. As the time constant of a low-pass filter also defines its response time, Allan deviation values for $1/\tau$ below the system filtering bandwidth become of no practical value.

Next, the fractional frequency noise spectral densities in terms of the resulting Allan deviations are discussed for different frequency tracking schemes.

5.2.2.1 Open-Loop Scheme (OLS)

The open loop frequency tracking scheme is schematically depicted in Fig. 5.7b. The mechanical resonator is driven by an external signal generator at $\omega = \omega_0$. The signal is branching into two arms: a reference arm and an arm that passes through the resonator where the signal experiences a phase shift. The phase difference between the two signals is then determined with an IQ (in-phase and quadrature component) demodulation. The resulting frequency change can now readily be calculated by multiplying the measured phase difference with the phase slope $1/\tau_R$, as schematically depicted in Fig. 5.8.

Assuming a noise-free signal generator, the steady-state transfer function for the thermomechanical ($S_{\varphi_{th}}$) and transduction ($S_{\varphi_{tr}}$) noise that are required to calculate $S_y(\omega)$ (5.39) is given by [4]

$$H_{\varphi_{th}}(i\omega) = H_R(i\omega)H_L(i\omega)$$

$$H_{\varphi_{tr}}(i\omega) = H_L(i\omega) \tag{5.44}$$

where $H_R(i\omega)$ & $H_L(i\omega)$ are low-pass filters. First the thermomechanical phase noise is low-pass filtered by the resonator itself with the transfer function [4]:

$$H_R(i\omega) = \frac{1}{1 + i\omega\tau_R} \tag{5.45}$$

with the resonator's time constant $\tau_R = 2Q/\omega_0$. Both the thermomechanical and transduction noises are filtered by the freely selectable low-pass filter of the IQ demodulator:

$$H_L(i\omega) = \frac{1}{1 + i\omega\tau_L}, \tag{5.46}$$

with the filter time constant τ_L.

With the given transfer functions (5.44), it is possible to numerically calculate the resulting Allan deviation (5.31) for different filter time constants τ_R and τ_L, as shown in Fig. 5.11.

The response time of such an open-loop frequency tracking scheme is ultimately limited by the time constant of the resonator itself, as can be seen from the presence of $H_R(i\omega)$ in $H_{\varphi_{th}}(i\omega)$ (5.44). A common situation is that the time constant of the resonator is shorter than the time constant of the demodulation filter ($\tau_R < \tau_L$). In other words, the resonance linewidth $\Gamma = \omega_0/Q = 2/\tau_R$ is larger than the filter bandwidth, and the system response time is limited by τ_L. In this case, the thermomechanical noise can be considered white with a constant value $S_{z_{th}}$ (5.11).

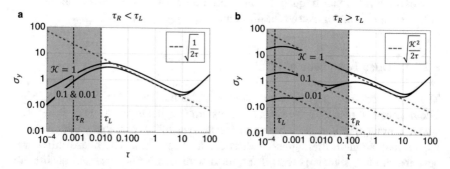

Fig. 5.11 Allan deviation (5.31) of the open-loop scheme calculated based on the transfer functions (5.44) for different time constant constellations and transduction noise levels. The calculations were performed for a normalized spectral density of the relative frequency fluctuation $S_{y_n}(\omega) = |H_{\varphi_{th}}(i\omega)|^2 + \mathcal{K}^2|H_{\varphi_{tr}}(i\omega)|^2$ (5.39). The sharp increase of σ_y for long integration time is due to frequency drift, which was randomly chosen. Integration times shorter than the system response time are shaded gray as they are of no practical use

The calculated corresponding Allan deviations for varying levels of transduction noise are plotted in Fig. 5.11a. It shows that for the entire practical range of integration times $\tau > \tau_L$, $\sigma_y(\tau)$ follows a the thermomechanical noise limit shown as red dashed line with a slope of $1/\sqrt{\tau}$. When transduction noise becomes equal in magnitude to thermomechanical noise ($\mathcal{K} = 1$), the Allan deviation starts to worsen.

Nanomechanical resonators can have very high quality factors Q, e.g., due to dissipation dilution as discussed in Sect. 3.3.3 on page 92. Such highly coherent resonators come with very narrow resonance peaks $\Gamma = \omega_0/Q = 2/\tau_R$, with values in the mHz range, and hence relatively long time constants τ_R. In this case, the demodulator filter time constant is typically shorter than the resonator time constant ($\tau_L < \tau_R$). As a result, the thermomechanical noise is no longer constant and takes the shape of a Lorentzian (1.34 on page 10) with a characteristic of a low-pass filter with the transfer function $H_R(i\omega)$ (5.45). The resulting Allan deviations for this scenario are shown in Fig. 5.11b. The resulting Allan deviation curves can feature a local minimum for integrations times $\tau < \tau_R$, which has been experimentally observed [17]. These minimal σ_y values are deceptive. Since these integration times are shorter than the response time, these Allan deviation values are of no practical use and have to be dismissed when calculating a sensor's sensitivity.

5.2.2.2   Phase-Locked Loop (PLL)

A phase-locked loop, in its broadest definition, is a control system that balances its operating point as a function of the difference between the phase of the output signal and its comparison with a given set point. Phase-locked loops are widely used, e.g., in telecommunications, radios, and computers. They can be used to recover a signal from a noisy communication channel, demodulate a signal, synthesize harmonic signals at multiples of an input frequency, etc.

In particular for the case of resonant nanomechanical sensors, the typical configuration is schematized in Fig. 5.9, and it has been used by many different groups in different variations [25–28]. The PLL scheme is similar to the OLS discussed in the previous subsection. However, instead of driving the resonator with a fixed frequency signal, the driving frequency is controlled in a negative feedback loop. A voltage-controlled oscillator (VCO) outputs a signal that splits into two branches: the reference arm and the measurement arm that passes through the resonator and hence observing a phase shift. The phase difference between the reference and resonator response is then determined by a IQ demodulator/phase detector, which typically is done with a lock-in amplifier. The resulting phase difference is then compared to the phase set point by subtracting it from the phase difference. The resulting phase error signal is the input for a *proportional-integral*/indexproportional-integral (PI) controller that creates a negative feedback and adjusts the frequency of the VCO to match the frequency of the resonator. The control signal from the PI control unit represents the wanted frequency shift $\Delta\omega_r$.

The derivation of the system's transfer functions for the phase noise is not as straightforward as in the OLS case. The steady-state transfer function for the thermomechanical and transduction phase noise that are required to calculate $S_y(\omega)$ (5.39) is given by [4]

$$H_{\varphi_{th}}(i\omega) = \frac{\left(i\omega K_p + K_i\right) H_L(i\omega)}{i\omega/\tau_R - \omega^2 + \left(i\omega K_p + K_i\right) H_L(i\omega)}$$

$$H_{\varphi_{tr}}(i\omega) = \frac{1}{H_R(i\omega)} \frac{\left(i\omega K_p + K_i\right) H_L(i\omega)}{i\omega/\tau_R - \omega^2 + \left(i\omega K_p + K_i\right) H_L(i\omega)},$$

(5.47)

with the steady-state low-pass filter transfer functions $H_L(i\omega)$ and $H_R(i\omega)$ for the IQ demodulator (5.46) and the resonator (5.45), respectively. K_P and K_I are the proportional and integral feedback controller parameter, respectively. The optimal values for the two parameters are [22, 29]

$$K_p = \frac{1}{\tau_{PLL}}, \text{ and } K_i = \frac{1}{\tau_R \tau_{PLL}},$$

(5.48)

where τ_{PLL} is the resulting PLL loop time constant. With these two PI parameters, the PLL transfer functions (5.47) simplifies to

$$H_{\varphi_{th}}(i\omega) = \frac{H_L(i\omega)}{H_L(i\omega) + i\omega\tau_{PLL}}$$

$$H_{\varphi_{tr}}(i\omega) = \frac{1}{H_R(i\omega)} \frac{H_L(i\omega)}{H_L(i\omega) + i\omega\tau_{PLL}}.$$

(5.49)

While τ_R is determined by the resonator, τ_L and τ_{PLL} can be chosen freely, with the constrain that $\tau_L < \tau_{PLL}$. It is a key feature of the PLL compared to the OLS that the response time is not limited by τ_R of the resonator. Instead, the speed of the PLL scheme is determined by τ_{PLL} and can be chosen freely via the setting of the PI controller parameters K_P and K_I. Figure 5.12 shows a PLL's Allan deviations based on the transfer functions (5.49) for different filter time constants τ_L and PLL time constants τ_{PLL}.

In an optimal case, the response of the resonator is faster than the PI controller ($\tau_{PLL} > \tau_R$), which is shown in Fig. 5.12a. Here, the resulting Allan deviation follows the thermomechanical noise limit, given by the red dashed line. The Allan deviation only gets worse once the transduction noise becomes of the order of the thermomechanical noise ($\mathcal{K} = 1$).

Figure 5.12b shows the scenario when the PLL time constant is set to match the resonator's time constant ($\tau_{PLL} = \tau_R$). In that case, the resulting Allan deviations are equal to the Allan deviations calculated for the OLS ($\tau_R > \tau_L$), presented in Fig. 5.11b. As long as the integration time τ remains larger than the systems response time, which is determined equally by τ_R and τ_{PLL}, the resulting Allan deviation follows the thermomechanical limit.

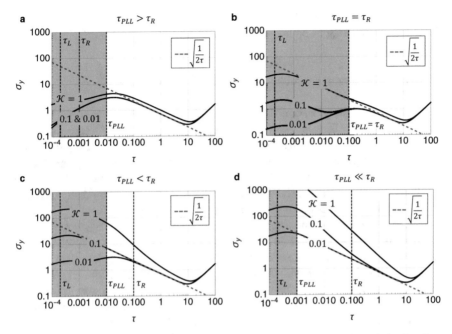

Fig. 5.12 Allan deviation (5.31) of the PLL scheme calculated based on the transfer functions (5.49) for different time constant constellations and transduction noise levels. The calculations were performed for a normalized spectral density of the relative frequency fluctuation $S_{y_n}(\omega) = |H_{\varphi_{th}}(i\omega)|^2 + \mathcal{K}^2|H_{\varphi_{tr}}(i\omega)|^2$ (5.39). The sharp increase of σ_y for long integration time is due to frequency drift, which was randomly chosen. Integration times shorter than the system response time are shaded gray as they are of no practical use

Finally, a very important effect of the PLL frequency tracking scheme is discussed, which is the scenario when the PLL response time is faster than the resonator response time ($\tau_{PLL} < \tau_R$). The corresponding Allan deviations are plotted in Fig. 5.12c and d. It is a key feature of a closed-loop frequency tracking scheme that the response time is not limited by the resonator's response time τ_R, as it is in the OLS. However, operating a PLL in the regime ($\tau_{PLL} < \tau_R$), which often is the case, e.g., with high-Q resonators, can result in a significant increase of the Allan deviation. Under the assumption $\tau_{PLL} \ll \tau_R$, the PLL bandwidth is a lot larger than the resonator linewidth, the transfer function (5.49) for the phase noise reduces to [4]

$$H_{\varphi_{tr}}(i\omega) \approx \frac{\tau_R}{\tau_{PLL}} H_L(i\omega) \ \ \text{for} \ \ \omega > 1/\tau_{PLL}. \tag{5.50}$$

This result implies that a PLL scheme amplifies transduction phase noise in the frequency range $1/\tau_L < \omega < 1/\tau_{PLL}$ with an amplification factor τ_R/τ_{PLL}. This noise amplification becomes significant when $\mathcal{K}(\tau_R/\tau_{PLL}) > 1$, as it can be seen in the calculated Allan deviations depicted in Fig. 5.12c and d. This effect has also been observed experimentally [17]. In conclusion, a PLL has serious restrictions for the frequency tracking of high-Q resonators where a fast response time is desired.

5.2.2.3   Self-Sustained Oscillator (SSO)

In the previous section, the concept of PLL is introduced and specifically how it can be used to monitor the frequency of nanomechanical resonators. As it was derived, using a PLL has certain limitations regarding the fast frequency tracking of high-Q resonator. An alternative is to send back to the resonator the amplitude signal after amplification and proper phase correction. This is typically called the *self-sustained oscillator* or the *positive feedback* technique. This has been tried by many different groups in different types of implementations[23, 24, 30–38].

Figure 5.9b shows a schematic for this technique. The gain in the amplifier (G) needs to be large enough as to compensate the losses within the resonator and transduction efficiency, whereas the phase shift (ϕ) needs to be adjusted so that the feedback signal is in phase with the velocity ($-\pi/2$) out of phase with respect to the amplitude, as schematically depicted in Fig. 5.8. An SSO needs a way to limit and control the oscillation amplitude. This is typically done with a nonlinear element in the oscillator loop, such as a saturating amplifier as shown in the schematic.

The equation governing the motion of a particular mode within a resonator is given by

$$m_{eff,n}\ddot{u}_n(t) + \Gamma_{eff,n}\dot{u}_n(t) + k_{eff,n}u_n(t) = G_m\dot{u}_n(t) + \xi_n(t), \tag{5.51}$$

where the transduction efficiency is hidden in the effective mechanical gain G_m, which is the gain in the saturating amplifier of Fig. 5.9b only with the transduction efficiency (both actuation and detection) embedded within. Equation (5.51) can be rewritten so that the conditions mentioned above become clearer:

$$m_{eff,n}\ddot{u}_n(t) + \left(\Gamma_{eff,n} - G_m\right)\dot{u}_n(t) + k_{eff,n}u_n(t) = \xi_n(t), \tag{5.52}$$

where it is now evident that when the condition $G_m > \Gamma_{eff,n}$ is held, the system becomes unstable, a self-sustained oscillation starts, and higher nonlinear terms need to be considered to actually saturate the motion. These two conditions are typically known as the *Barkhausen criteria* for oscillation, i.e., the gain in the feedback compensates the losses in the resonator and the phase is matched to the velocity.

Once the system is operating, one needs to measure the frequency with a frequency detector. This can be done in different ways, e.g., by a phase measurement with a lock-in amplifier or a frequency counter. In any case, the frequency detector has a given measurement time τ_L over which the frequency signal is averaged. Hence, a frequency detector has a low-pass filter characteristic with a transfer function $H_L(i\omega)$ (5.46).

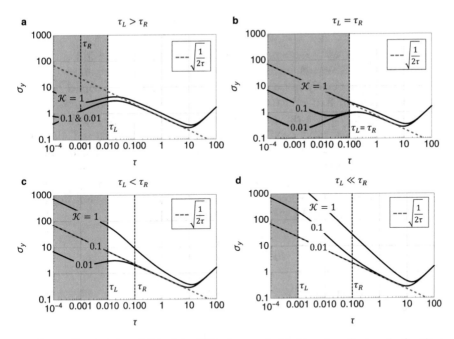

Fig. 5.13 Allan deviation (5.31) of the SSO scheme calculated based on the transfer functions (5.53) for different time constant constellations and transduction noise levels. The calculations were performed for a normalized spectral density of the relative frequency fluctuation $S_{y_n}(\omega) = |H_{\varphi_{th}}(i\omega)|^2 + \mathcal{K}^2|H_{\varphi_{tr}}(i\omega)|^2$ (5.39). The sharp increase of σ_y for long integration time is due to frequency drift, which was randomly chosen. Integration times shorter than the system response time are shaded gray as they are of no practical use

The transfer functions for the thermomechanical and transduction phase noise of a SSO are rather simple [39]

$$H_{\varphi_{th}}(i\omega) = H_L(i\omega)$$

$$H_{\varphi_{tr}}(i\omega) = \frac{H_L(i\omega)}{H_R(i\omega)}. \tag{5.53}$$

The resulting Allan deviations for a SSO for various resonator and filter time constants are plotted in Fig. 5.13. It is clear that the Allan deviations obtained with an SSO are equal to the ones with a PLL for all permitted integration times as shown in Fig. 5.12. In the case of negligible transduction noise, the Allan deviation follows the thermomechanical noise of the resonator with a slope $1/\sqrt{\tau}$. For substantial transduction noise and $\tau_R > \tau_L$, the Allan deviations become worse with a slope $1/\tau$. Also like a PLL, the SSO frequency tracking scheme responds instantaneously to frequency changes of the resonator [22], and the response time is determined by τ_L of the frequency detector.

5.2.3 Frequency Noise from Thermal Fluctuations

Besides the fundamental thermomechanical noise limit discussed in the previous section, statistical temperature fluctuations are a potentially significant source contributing to frequency noise. Temperature fluctuation noise comes from the statistical nature of heat exchange between the resonator and its surroundings. When in vacuum, the heat exchange with the environment happens either through phonons via thermal conduction or through photons via thermal radiation. The resulting noise produced by the quantum nature of these thermal fluctuations is shot noise. As a simplified first-order model, we will consider the resonator as a concentrated mass linked to the reservoir via a thermal conductance G. Based on the fluctuation-dissipation theorem, the corresponding white temperature fluctuation spectral density with units [K^2/Hz] is then given by [21, 40]

$$S_T = \frac{4k_B T^2}{G} H_{th}^2(\omega). \tag{5.54}$$

The nonzero heat capacity of the resonator acts as a low-pass filter with the transfer function

$$H_{th}(\omega) = \frac{1}{\sqrt{1 + (\omega\tau_{th})^2}}, \tag{5.55}$$

and the thermal time constant [41]

$$\tau_{th} = \frac{C_{th}}{G} \tag{5.56}$$

with the heat capacity C_{th} and thermal conductance G. Under vacuum conditions, the thermal conductance G comprises heat transfer through thermal conduction and radiation.

These white noise temperature fluctuations translate into frequency fluctuations leveraged by the resonator-specific responsivity to temperature $\mathscr{R}_T$ with units [1/K]. The resulting fractional frequency noise spectral density then becomes:

$$S_{y_{thf}}(\omega) = S_T \, [\mathscr{R}_T(\omega)]^2. \tag{5.57}$$

The resonator-specific responsivities to temperature are discussed in Sect. 6.2.2 on page 189.

Temperature fluctuations are the fundamental noise source that constitute the sensitivity limit of thermal infrared detectors [41]. Recently, evidence has been presented that frequency fluctuations can be the limiting noise source in silicon nitride drum resonators [42]. In Sect. 6.2.3 on page 194, the responsivity to absorbed power and the sensitivity of nanomechanical drumhead resonators are discussed with respect to temperature fluctuations.

References

1. Callen, H. B., & Welton, T. A. (1951). Irreversibility and generalized noise. *Physical Review, 83*(1), 34–40.
2. Nyquist, H. (1928). Thermal agitation of electric charge in conductors. *Physical Review, 32*(1), 110–113.
3. Albrecht, T. R., Grütter, P., Horne, D., & Rugar, D. (1991). Frequency modulation detection using high-Q cantilevers for enhanced force microscope sensitivity. *Journal of Applied Physics, 69*(2), 668.
4. Demir, A. (2021). Understanding fundamental trade-offs in nanomechanical resonant sensors. *Journal of Applied Physics, 129*(4), 044503.
5. Johnson, J. B. (1927). Thermal agitation of electricity in conductors. *Nature, 119*, 50–51.
6. Roukes, M. L., Freeman, M. R., Germain, R. S., Richardson, R. C., & Ketchen, M. B. (1985). Hot-electrons and energy-transport in metals at millikelvin temperatures. *Physical Review Letters, 55*(4), 422–425.
7. Campbell, N. (1910). The study of discontinuous phenomena. *Proceedings of the Cambridge Philosophical Society, 15*, 117–136.
8. Campbell, N. (1910). Discontinuities in light emission. *Proceedings of the Cambridge Philosophical Society, 15*, 310–328.
9. Mathieson, E. (1977). Derivation of noise formulas using campbells theorem. *American Journal of Physics, 45*(12), 1184–1186.
10. Schottky, W. (1918). Regarding spontaneous current fluctuation in different electricity conductors. *Annalen Der Physik, 57*(23), 541–567.
11. Beenakker, C. & Schonenberger, C. (2003). Quantum shot noise. *Physics Today, 56*(5), 37–42.
12. Halford, D. (1966). A general mechanical model of random perturbations which generate a noise spectral density law [f]alpha with reference to flicker noise law 1/[f]. *IEEE Transactions on Instrumentation and Measurement, 15*(4), 404.
13. Keshner, M. S. (1982). 1/f noise. *Proceedings of the IEEE, 70*(3), 212–218.
14. Hooge, F. N. (1976). 1/f noise. *Physica B & C, 83*(1), 14–23.
15. Milotti, E. (1995). Linear processes that produce 1/f or flicker noise. *Physical Review E, 51*(4), 3087.
16. Motchenbacher, C. D., Connelly, J. A., & Motchenbacher, C. D. (1993). *Low-noise electronic system design*. New York: Wiley.
17. Sadeghi, P., Demir, A., Villanueva, L. G., Kähler, H., & Schmid, S. (2020). Frequency fluctuations in nanomechanical silicon nitride string resonators. *Physical Review B, 102*(21), 214106.
18. Allan, D. W., Machlan, H. E., & Gray, J. E. (1972). National bureau of standards atomic time scales—generation, dissemination, stability, and accuracy. *IEEE Transactions on Instrumentation and Measurement, 21*(4), 388.
19. Walls, F. L., & Allan, D. W. (1986). Measurements of frequency stability. *Proceedings of the IEEE, 74*(1), 162–168.
20. Allan, D. A., Ashby, N., & Hodge, C. C. (1997). The science of timekeeping. Report, Hewlett Packard.
21. Cleland, A. N., & Roukes, M. L. (2002). Noise processes in nanomechanical resonators. *Journal of Applied Physics, 92*(5), 2758–2769.
22. Demir, A., & Selim Hanay, M. (2019). Fundamental sensitivity limitations of nanomechanical resonant sensors due to thermomechanical noise. *IEEE Sensors Journal, 20*(4), 1947–1961.
23. Villanueva, L. G., Kenig, E., Karabalin, R. B., Matheny, M. H., Lifshitz, R., Cross, M. C., & Roukes, M. L. (2013). Surpassing fundamental limits of oscillators using nonlinear resonators. *Physical Review Letters, 110*(17), 177208.
24. Kenig, E., Cross, M. C., Villanueva, L. G., Karabalin, R. B., Matheny, M. H., Lifshitz, R., & Roukes, M. L. (2012). Optimal operating points of oscillators using nonlinear resonators. *Physical Review E, 86*(5), 056207.

25. Naik, A. K., Hanay, M. S., Hiebert, W. K., Feng, X. L., & Roukes, M. L. (2009). Towards single-molecule nanomechanical mass spectrometry. *Nature Nanotechnology, 4*(7), 445–450.
26. Yang, Y. T., Callegari, C., Feng, X. L., Ekinci, K. L., & Roukes, M. L. (2006). Zeptogram-scale nanomechanical mass sensing. *Nano Letters, 6*(4), 583–586.
27. Yang, Y. T., Callegari, C., Feng, X. L., & Roukes, M. L. (2011). Surface adsorbate fluctuations and noise in nanoelectromechanical systems. *Nano Letters, 11*(4), 1753–1759.
28. Chaste, J., Eichler, A., Moser, J., Ceballos, G., Rurali, R., & Bachtold, A. (2012). A nanomechanical mass sensor with yoctogram resolution. *Nature Nanotechnology, 7*(5):300–303.
29. Olcum, S., Cermak, N., Wasserman, S. C., & Manalis, S. R. (2015). High-speed multiple-mode mass-sensing resolves dynamic nanoscale mass distributions. *Nature Communications, 6*(1), 1–8.
30. Schmid, S., Senn, P., & Hierold, C. (2008). Electrostatically actuated nonconductive polymer microresonators in gaseous and aqueous environment. *Sensors and Actuators: A. Physical, 145–146*, 442–448.
31. Feng, X. L., White, C. J. , Hajimiri, A., & Roukes, M. L. (2008). A self-sustaining ultrahigh-frequency nanoelectromechanical oscillator. *Nature Nanotechnology, 3*(6), 342–346.
32. Ayari, A., Vincent, P., Perisanu, S., Choueib, M., Gouttenoire, V., Bechelany, M., Cornu, D., & Purcell, S. T. (2007). Self-oscillations in field emission nanowire mechanical resonators: A nanometric DC-AC conversion. *Nano Letters, 7*(8), 2252–2257.
33. Vincent, P., Perisanu, S., Ayari, A., Choueib, M., Gouttenoire, V., Bechelany, M., Brioude, A., Cornu, D., & Purcell, S. T. (2007). Driving self-sustained vibrations of nanowires with a constant electron beam. *Physical Review B, 76*(8), 085435.
34. Villanueva, L. G., Karabalin, R. B., Matheny, M. H., Kenig, E., Cross, M. C., & Roukes, M. L. (2011). A nanoscale parametric feedback oscillator. *Nano Letters, 11*(11), 5054–5059.
35. Chen, C. Y., Lee, S., Deshpande, V. V., Lee, G. H., Lekas, M., Shepard, K., & Hone, J. (2013). Graphene mechanical oscillators with tunable frequency. *Nature Nanotechnology, 8*(12), 923–927.
36. Zuo, C. J., Van der Spiegel, J., & Piazza, G. (2011). Dual-mode resonator and switchless reconfigurable oscillator based on piezoelectric ALN MEMS technology. *IEEE Transactions on Electron Devices, 58*(10), 3599–3603.
37. Zuo, C. J., Van der Spiegel, J., & Piazza, G. (2010). 1.05-GHZ CMOS oscillator based on lateral-field-excited piezoelectric ALN contour-mode MEMS resonators. *IEEE Transactions on Ultrasonics Ferroelectrics and Frequency Control, 57*(1), 82–87.
38. Hui, Y., Nan, T. X., Sun, N. X., & Rinaldi, M. (2015). High resolution magnetometer based on a high frequency magnetoelectric mems-cmos oscillator. *Journal of Microelectromechanical Systems, 24*(1), 134–143.
39. Bečić, H., Demir, A., Steurer, J., Luhmann, N., & Schmid, S. Resonance frequency tracking schemes for micro- and nanomechanical resonators. Manuscript in preparation at the time of writing.
40. Vig, J. R., & Kim, Y. (1999). Noise in microelectromechanical system resonators. *IEEE transactions on Ultrasonics, Ferroelectrics, and Frequency Control, 46*(6), 1558–1565.
41. Kruse, P. W., McGlauchlin, L. D., & McQuistan, R. B. (1962). *Elements of infrared technology: Generation, transmission and detection.* New York: Wiley.
42. Zhang, C., & St-Gelais, R. (2022). Demonstration of frequency stability limited by thermal fluctuation noise in silicon nitride nanomechanical resonators. arXiv preprint arXiv:2211.12651.

Chapter 6
Responsivity and Sensitivity

The rational to develop nanomechanical resonators is to achieve unique sensors with exceptional sensitivities. Nanomechanical resonators are known for their exceptional force sensitivity [1, 2]. The response $z(t) = z_0 \cos(\omega t - \varphi)$ of a resonator to an external force $F(t) = F_0 \cos(\omega t)$ has been discussed in detail in Sect. 1.1.3 on page 5. As discussed, if the frequency of the force is a lot smaller than the resonance frequency (quasi-static regime), the system essentially responds like a simple linear spring with a quasi-static displacement $z_s = F_0/k$. If the frequency of the force matches the eigenfrequency of the system, according to (1.22), the response gets amplified by the quality factor $z_0 = Q \cdot F_0/k$. In contrast to the quasi-static regime where a force change causes an instantaneous amplitude response, the response time at resonance is determined by the resonators' time constant $\tau_R = 2Q/\omega_0$, as discussed in Sect. 1.1.3.2 on page 10. In Sect. 1.4 on page 25, it was further shown that parametrically pumping a resonator can cause a parametric amplification of small forced vibrations. The force sensitivity when measured via the vibrational amplitude of a nanomechanical resonator is ultimately determined by the force noise spectral density (5.10 on page 148, discussed in 5.1.2).

In contrast to the vibrational response to an external force, in this chapter the response of a linear resonator to changes of its parameters is discussed. Examining the equation of motion of a linear driven resonator (1.1)

$$m\ddot{z}(t) + c\dot{z}(t) + kz(t) = F(t) \tag{6.1}$$

readily reveals all parameters, which are the mass m, damping coefficient c, and spring constant k, whose change induces a change in the system response. The resonator response can be characterized by its eigenfrequency ω_0, quality factor Q, and amplitude response z_0, as visualized in Fig. 6.1. For a slightly damped system, all three response characteristics are a function of the three system parameters (see Sect. 1.1 on page 2):

© The Author(s), under exclusive license to Springer Nature Switzerland AG 2023
S. Schmid et al., *Fundamentals of Nanomechanical Resonators*,
https://doi.org/10.1007/978-3-031-29628-4_6

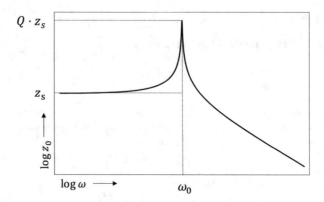

Fig. 6.1 Key characteristics z_s, Q, and ω_0 of the response of a driven linear resonator under the assumption of slight damping

$$\omega_r \approx \omega_0 = \sqrt{\frac{k}{m}}$$

$$Q \approx \frac{\sqrt{k\,m}}{c}$$

(6.2)

$$|z_0| = \begin{cases} z_s = \frac{F_0}{k}, & \omega \ll \omega_0 \\ z_r = Q \cdot z_s = \frac{F_0}{c}\sqrt{\frac{m}{k}}, & \omega = \omega_0. \end{cases}$$

A change of any of these parameters can be detected via these observable quantities. However, not all observable quantities are equally applicable.

This chapter is organized by the response to changes of each parameter, with the exception of the damping coefficient. Applications relying on the detection of c are rare, such as the determination of loss tangent $\tan\delta$ of a viscoelastic material [3, 4] or the study surface oxidation [5]. However, in these applications, the actual parameter of interest is the quality factor itself ($\tan\delta = Q^{-1}$) and not the coefficient of damping alone, as discussed in Sect. 3.3 on page 81.

For sensing changes of mass and stiffness, the eigenfrequency ω_0 is the most robust and frequently used observable, and it will be the main focus of this chapter. In contrast to the vibrational amplitude z_r, the eigenfrequency is independent of the driving force and hence of the transduction scheme. And in particular for mass sensing, z_r is sensitive to the mass momentum and timing of absorption with respect to the oscillation of the nanomechanical resonator. While detecting changes of mass leads to obvious applications, e.g., for mass spectrometry [6] or nanoparticle detection [7], detecting a change of stiffness can be used, e.g., to detect temperature changes or force gradients, as will be described in this chapter.

It is noteworthy that a change of a parameter has an immediate effect on the resonators' eigenfrequency. It is possible to follow instantaneous parameter changes with high fidelity with the help of closed-loop oscillator circuits that use the nanomechanical resonator as frequency determining element, as introduced in Sect. 5.2 on page 158.

Before the discussion of the resonator response to each parameter, the sensor terminology used in this book is briefly introduced.

- *Responsivity* ($\mathcal{R}_\xi$) is the fractional slope of the sensor output as a function of an input quantity ξ. The responsivity is constant over the linear range of the sensor. In the case of a nanomechanical resonator, sensor output is a change of one of the three observable variables (6.2). ξ is any quantity that ultimately affects at least one of the elementary input parameters mass, damping, and stiffness. In the case of the eigenfrequency as observable for sensing a change in ξ, the responsivity can be defined as

$$\mathcal{R}_\xi = \frac{\partial \omega_0}{\partial \xi} \frac{1}{\omega_0}. \tag{6.3}$$

The responsivity leverages a small fractional eigenfrequency change $y(t) = \Delta\omega_0(t)/\omega_0$ into a signal corresponding to the specific input quantity change $\Delta\xi(t)$:

$$\Delta\xi(t) = \frac{\Delta\omega_0(t)}{\omega_0} \mathcal{R}_\xi^{-1} = \frac{y(t)}{\mathcal{R}_\xi}. \tag{6.4}$$

- *Sensitivity* is the smallest detectable change of the input parameter with a signal-to-noise ratio of one. With the eigenfrequency as observer quantity, the sensitivity for a certain integration time τ can be calculated from the corresponding Allan deviation $\sigma_y(\tau)$, as discussed in Sect. 5.2.1 on page 160:

$$\Delta\xi_{min}(\tau) = \frac{\sigma_y(\tau)}{\mathcal{R}_\xi}. \tag{6.5}$$

Alternatively, the sensitivity can be expressed more generally as a power spectral density:

$$S_\xi(\omega) = \frac{S_y(\omega)}{\mathcal{R}_\xi^2} \tag{6.6}$$

with the fractional-frequency noise spectral density $S_y(\omega)$. $S_y(\omega)$ is the sum of uncorrelated fractional-frequency noise sources, e.g., from thermomechanical noise or thermal fluctuations, as discussed in Sect. 5.2 on page 158. According to (5.33), for white fractional frequency noise, the sensitivity can also be expressed as a function of the Allan deviation for a specific integration time τ

$$S_\xi(\omega) = \frac{\sigma_y^2(\tau)2\tau}{\mathcal{R}_\xi^2}. \tag{6.7}$$

6.1 Response to Change of Mass ($\mathscr{R}_m$)

The probably most prominent sensor application for nanomechanical resonators is the detection of small masses. As can be seen from the list of observables (6.2), changes in mass influence both the resonance frequency and the quality factor. However, since it's a lot easier to accurately determine the former, changes of mass are commonly detected via the frequency response.

Assuming that the nanomechanical resonator is only slightly damped, then the resonance frequency can be approximated by the eigenfrequency ω_0 (1.20). In that case, according to (6.3), the responsivity is given by the fractional slope of the frequency at the initial effective mass m_{eff} of the resonator:

$$\mathscr{R}_m = \frac{1}{\omega_0(m_{eff})}\left.\frac{\partial\omega_0}{\partial m}\right|_{m=m_{eff}} = -\frac{1}{2m_{eff}}. \tag{6.8}$$

Obviously, in order to obtain a high responsivity, the resonator must have a low effective mass m_{eff}. This is one of the main rational behind the development of nanomechanical resonators which have a mass typically below 1 ng.

Assuming that the change of mass Δm is small, the resulting response can be linearized, as schematically depicted in Fig. 6.2, which according to (6.4) gives:

$$\Delta m(t) = \mathscr{R}_m^{-1}y(t). \tag{6.9}$$

The mass loading can happen in many ways. Here two common scenarios are discussed, namely, point mass loading and distributed mass loading.

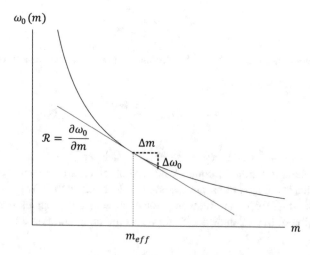

Fig. 6.2 Eigenfrequency of a linear resonator as a function of mass m. The red line represents the slope at the effective mass m_{eff}, which is the responsivity $\mathscr{R}$

Fig. 6.3 Schematic of a
cantilever with a single bead,
having the mass Δm,
positioned at $x_{\Delta m}$

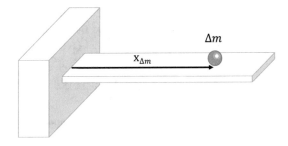

6.1.1   Response to Point Mass

In the previous section, the mass of adsorbed molecules is assumed to be distributed
uniformly over the resonator surface. This approach is not viable if single molecules
or particles are to be measured, since the change in frequency is dependent not only
on the mass of the attached particle but also on the landing position on the resonator
[8]. This is due to the shape of the vibrational modes. The areas of the beam with a
large vibrational amplitude are areas where an added mass will gain a high kinetic
energy and thereby change the resonant frequency considerably compared with the
nodal points where a point mass remains unmoved.

Considering a cantilever with the effective mass m_{eff} loaded with a point mass
Δm positioned at $x_{\Delta m}$, as schematically depicted in Fig. 6.3. If the mass load
is much smaller than the resonator mass, $\Delta m \ll m_{eff}$, the cantilever mode
shape will not change significantly, and the resonant frequency of such a system
can be accurately estimated using an energy approach. According to Rayleigh's
method (2.1), the maximum of the kinetic energy, $(W_{kin}$, equals the maximum of
the strain energy W_{strain}, at resonance. In the case of a resonator with an additional
point mass with the kinetic energy $W_{kin,\Delta m}$, the total energy balance becomes [9]:

$$\max{(W_{strain})} = \max{(W_{kin})} + \max{(W_{kin,\Delta m})}. \tag{6.10}$$

The kinetic energy of the beam can be calculated by considering the kinetic
energy of an infinitesimally small piece of the beam of mass dm and length dx:

$$dW_{kin} = \frac{1}{2}dm\left(\frac{\partial U_n(x,t)}{\partial t}\right)^2 = \frac{1}{2}A\rho\left(\frac{\partial U_n(x,t)}{\partial t}\right)^2 dx, \tag{6.11}$$

where A and ρ are the beam cross section and mass density, respectively. Integrating (6.11) it over the entire beam length L, considering the displacement function
$U_n(x,t) = u_{0,n}\cos{(\omega_n t)}\phi_n(x)$ of a normal mode n, results in

$$\max{(W_{kin})} = \frac{1}{2}A\rho\omega_{n,\Delta m}^2 u_{0,n}^2 \int_0^L \phi_n^2(x)dx = \frac{1}{2}u_{0,n}^2\omega_{n,\Delta m}^2 m_{eff}, \tag{6.12}$$

where $\omega_{n,\Delta m}$ is the frequency of motion with the additional mass, $\phi_n(x)$ is the normalized modeshape function of the n's mode and $u_{0,n}$ is the modal amplitude of each mode, and the effective mass (2.124)

$$m_{eff} = \rho A \int_0^L \phi_n^2(x)dx = m_0 \frac{1}{L} \int_0^L \phi_n^2(x)dx. \tag{6.13}$$

The maximum kinetic energy of the added point mass Δm at $x_{\Delta m}$ can readily be calculated as

$$\max(W_{kin,\Delta m}) = \frac{1}{2}\Delta m \omega_{n,\Delta m}^2 u_{0,n}^2 \phi_n^2(x_{\Delta m}). \tag{6.14}$$

Assuming that neither the mode shape nor the moment of inertial will change significantly due to the added mass, the maximum strain energy in the cantilever is approximately equal to the kinetic energy of the cantilever without the added mass, that is, with eigenfrequency ω_n instead of $\omega_{n,\Delta m}$:

$$\max(W_{strain}) \approx \frac{1}{2}u_{0,n}^2 \omega_n^2 m_{eff}. \tag{6.15}$$

After inserting (6.12), (6.14), and (6.15) in (6.10), the eigenfrequency of the beam with the added mass then becomes:

$$\omega_{n,\Delta m}^2 = \omega_n^2 \left(1 + \frac{\Delta m}{m_{eff}}\phi_n^2(x_{\Delta m})\right)^{-1}. \tag{6.16}$$

Since the additional mass is assumed to be a lot smaller than the cantilever mass, (6.16) can be simplified by a Taylor series approximation:

$$\omega_{n,\Delta m} \approx \omega_n \left(1 - \frac{1}{2}\frac{\Delta m}{m_{eff}}\phi_n^2(x_{\Delta m})\right). \tag{6.17}$$

The point mass responsivity for a specific mode n according to (6.8) then becomes [6, 9, 10]:

$$\mathscr{R}_{\Delta m} = -\frac{1}{2m_{eff}}\phi_n^2(x_{\Delta m}). \tag{6.18}$$

From (6.18) it is obvious that the position of the mass has to be known prior to the calculation of the mass itself. Hence, there are two unknowns in total, namely, the position $x_{\Delta m}$ and the mass Δm. Both unknowns can be obtained by creating a system of equations with (6.18) based on at least two vibrational modes n.

6.1.1.1 Response of Strings to Point Mass

The single mass detection with strings is discussed first. The sinusoidal modeshape function allows for a simple analysis [11]. With the modeshape function of a string (2.49) with length L

$$\phi_n(x) = \sin\left(\frac{n\pi x}{L}\right) \tag{6.19}$$

where n is the mode number, and the effective mass (see Table 2.4) $m_{eff} = \frac{1}{2}m_0$, with the total string mass $m_0 = \rho A L$, the point mass Δm positioned at $x_{\Delta m}$ can readily be found from the responsivity (6.18):

$$\Delta m = y\mathscr{R}_{\Delta m}^{-1} = -\frac{\Delta\omega_n}{\omega_n}m_0\sin^{-2}\left(\frac{n\pi}{L}x_{\Delta m}\right). \tag{6.20}$$

The point mass Δm and its position $x_{\Delta m}$ are the unknowns of a defined second-order system of equations based on (6.20) for the first two bending modes. For the first bending mode ($n = 1$), (6.20) can be solved for the position:

$$x_{\Delta m} = \frac{L}{\pi}\arcsin\sqrt{-\frac{m_0}{\Delta m}\frac{\Delta\omega_1}{\omega_1}}. \tag{6.21}$$

The absolute string displacement is symmetrical, and it does not make a difference on which half side a point mass is added to the string. The resulting frequency shift is the same. Therefore, the positions resulting from (6.21) have only values from 0 to $L/2$.

The mass ratio of the point mass versus string mass can now be obtained by substituting (6.21) in (6.20) for the second mode ($n = 2$). By using the identity $(\sin(2\arcsin x) = 2x\sqrt{1 - x^2})$, a simple term for the relative point mass can be calculated:

$$\frac{\Delta m}{m_0} = \frac{4\left(\frac{\Delta\omega_1}{\omega_1}\right)^2}{\frac{\Delta\omega_2}{\omega_2} - 4\frac{\Delta\omega_1}{\omega_1}}. \tag{6.22}$$

The used identity holds only if $|x| < 1$. It can easily be shown that this condition is fulfilled if $x_{\Delta m} < L/2$ which is in agreement with the before-mentioned symmetry of the string vibration.

With (6.22), the relative particle position (6.21) can be readily calculated:

$$\frac{x_{\Delta m}}{L} = \frac{1}{\pi}\arcsin\sqrt{1 - \frac{1}{4}\frac{\Delta\omega_2}{\omega_2}\frac{\omega_1}{\Delta\omega_1}}. \tag{6.23}$$

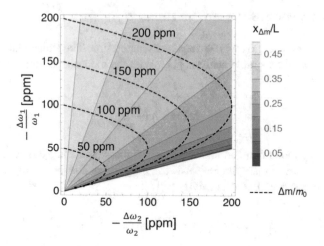

Fig. 6.4 Visualization of the particle location (6.23) and corresponding particle mass (6.22) as a function of the relative frequency shifts of the first and second bending mode

Both the relative position (6.23) and the relative mass change (6.22) are plotted in Fig. 6.4 as a function of the relative frequency shifts for the first and second mode. It can be seen that the resulting masses are collapsing for particle positions close to the clamping site. The real measured relative frequency changes have a certain imprecision, as discussed in detail in Sect. 5.2 on page 158. Hence, masses that land close to the clamping, where the dashed mass lines collapse, will produce a highly imprecise mass response.

Assuming a single uncertainty for the relative frequency shift of both modes σ_{ω_0}, the uncertainty of the relative mass change (6.22) is given by the Gaussian error propagation for uncorrelated variables:

$$\sigma_m = \sigma_{\omega_0} \sqrt{\left(\frac{\partial \frac{\Delta m}{m_0}}{\partial \frac{\Delta \omega_1}{\omega_1}}\right)^2 + \left(\frac{\partial \frac{\Delta m}{m_0}}{\partial \frac{\Delta \omega_2}{\omega_2}}\right)^2}. \tag{6.24}$$

With (6.22) and (6.23), the mass uncertainty (6.24) can be transformed into a function of the mass position $x_{\Delta m}$:

$$\frac{\sigma_m}{\sigma_{\omega_0}} = \frac{1}{4} \sqrt{9 + 8 \frac{\cos\left(4\pi \frac{x_{\Delta m}}{L}\right)}{\left(\sin\left(\pi \frac{x_{\Delta m}}{L}\right)\right)^8}}, \tag{6.25}$$

which is plotted in Fig. 6.5a. It shows that there are two regimes. For a position in the string center ($0.25 < x_{\Delta m}/L < 0.75$), an error in the frequency measurement translates directly into an error in the calculated mass with a factor of unity. For a position close to the clamping sites ($0 < x_{\Delta m}/L < 0.25$ and $0.75 < x_{\Delta m}/L <$

Fig. 6.5 (**a**) Ratio of mass to
frequency uncertainty
$\sigma_m/\sigma_{\omega_0}$ (6.25) as a function
of the relative mass position
$x_{\Delta m}/L$ along the length L of
the resonator. (**b**) Modeshape
plots of the first two modes
with the active area available
for mass sensing marked red

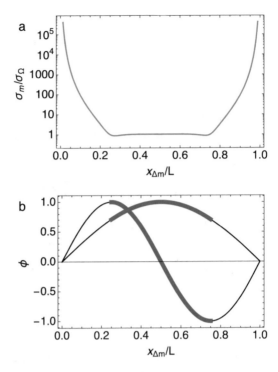

Fig. 6.5 (**a**) Ratio of mass to frequency uncertainty $\sigma_m/\sigma_{\omega_0}$ (6.25) as a function of the relative mass position $x_{\Delta m}/L$ along the length L of the resonator. (**b**) Modeshape plots of the first two modes with the active area available for mass sensing marked red

1), an error in the frequency measurement is amplifying the mass uncertainty. Therefore, in single mass detection, masses that land at the ends of the resonator are typically dismissed, and only masses that land in the center ($0.25 < x_{\Delta m}/L < 0.75$) are considered for the measurement, that is, only 50% of the sensor surface are used [6, 11], as depicted in Fig. 6.5b.

6.1.1.2 Response of Beams to Point Mass

The more complex modeshape function of beams (cantilevers and bridges) is more complicated and does not allow the derivation of a closed-form solution, as done for strings in the previous section. The point mass responsivity according to (6.18) of cantilevers and bridges can be calculated by means of the respective normalized modeshape functions (2.37) and (2.46) as derived in Sect. 2.2, and the respective effective masses (see Table 2.4 on page 60) $m_{eff} = \frac{1}{4}m_0$ and $m_{eff} \approx \frac{1}{2}m_0$, respectively.

If a single point mass is adhering to the cantilever, and the change in resonant frequency of several modes is measured, both position and mass can be calculated. For this purpose, based on the resonant frequency changes, two column vectors are constructed:

$$\overline{\phi}(x_{\Delta m}) = \begin{bmatrix} \phi_1^2(x_{\Delta m}) \\ \phi_2^2(x_{\Delta m}) \\ \vdots \\ \phi_n^2(x_{\Delta m}) \end{bmatrix}, \quad \overline{y} = \begin{bmatrix} \frac{\Delta \omega_1}{\omega_1} \\ \frac{\Delta \omega_2}{\omega_2} \\ \vdots \\ \frac{\Delta \omega_n}{\omega_n} \end{bmatrix}. \tag{6.26}$$

The size of the vectors is determined by the number of measured vibrational modes, with a minimum of two modes required to determine the position. Equation (6.18) can now be written as

$$\overline{y} = -\frac{\Delta m}{2 m_{eff}} \, \overline{\phi}(x_{\Delta m}). \tag{6.27}$$

This problem cannot be solved numerically. Therefore first, the particle positions are recovered by creating a term that can be minimized numerically. This is done by creating a normalized scalar product which approaches unity when $\overline{\phi}(x_{\Delta m})$ and $\overline{y}$ become collinear, that is, when (6.27) is satisfied:

$$\psi(x_{\Delta m}) = 1 - \frac{\left| \overline{\phi}(x_{\Delta m}) \cdot \overline{y} \right|}{\left\| \overline{\phi}(x_{\Delta m}) \right\| \cdot \left\| \overline{y} \right\|}. \tag{6.28}$$

The position $x_{\Delta m}$ can now be found by numerically minimizing ψ. From the computed position, the relative mass change induced by the individual particles can then readily be obtained by

$$\Delta m = -2 m_{eff} \frac{\left\| \overline{y} \right\|}{\left\| \overline{\phi}(x_{\Delta m}) \right\|}. \tag{6.29}$$

This minimization technique has been extended to detect multiple masses on a single cantilever by measuring the frequency shifts of multiple resonance modes [12].

A different approach to calculate the mass and position is by mapping the plane of the frequency shifts to the plane of the mass and position (as plotted for a bridge in Fig. 6.6) via a bivariate transformation by treating the variables as independent probability density functions [6].

The latest development in the field of nanomechanical mass sensing is the detection of the mass density distribution of a point mass, which allows to gather an inertial image with non-diffraction-limited resolution [13].

6.1.2   Response to Distributed Mass

A common way of using a nanomechanical resonator as mass sensor, e.g., for nanoparticle concentration measurements [7, 14], detection of flux of atoms [15], or gas sensing by coating the top surface of the resonator with a sensitive layer that

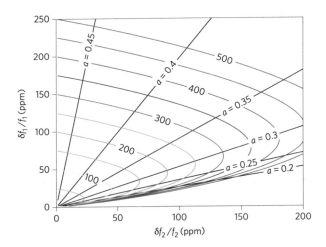

Fig. 6.6 Visualisation of the particle location (6.23) and corresponding particle mass (6.22) as a function of the relative frequency shifts of the first and second bending mode of a nanomechanical bridge. The black straight lines represent the location $x_{\Delta m} = a$. Colored lines represent the relative mass change $\Delta m/m_{eff}$ in ppm (Reprinted from [6])

absorbs the gas to be detected [16], as schematically depicted in Fig. 6.7. The gas is then absorbed in the sensitive layer which increases the total resonator mass. Under the assumption that the additional mass does not alter the stiffness of the resonator, the absorbed mass can be interpreted as an increase in mass density of the resonator.

In the case of an evenly distributed mass over the sensor surface area, the figure of merit becomes the mass sensitivity per surface area A_S, instead of the total mass sensitivity. The distributed mass load for a small additional mass can then be calculated by according to (6.8)

$$\frac{\Delta m}{A_s}(t) = \frac{y(t)}{A_s \mathscr{R}_{\Delta m}} = \frac{y(t)}{\mathscr{R}_{\Delta m/A_s}} \tag{6.30}$$

with the distributed mass responsivity

$$\mathscr{R}_{\Delta m/A_s} = -\frac{A_s}{2m_0}. \tag{6.31}$$

If the entire resonator is covered with mass, the responsivity reduces to

$$\mathscr{R}_{\Delta m/A_s} = -\frac{1}{2h\rho} \tag{6.32}$$

with the resonator mass density ρ and thickness h. When measuring distributed mass, the responsivity is maximal for a thin resonant structure with minimal density. The lateral dimensions are irrelevant as long as the entire surface experiences an

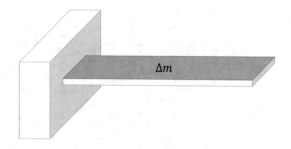

Fig. 6.7 Schematic of a cantilever with an additional mass Δm evenly distributed over the entire surface area

even mass loading. Hence, thin membrane resonators are optimal designs for such applications.

6.2 Response to Change of Effective Spring Constant

The spring constant is an input parameter that appears in all three resonator characteristics (6.2), and can be detected via the eigenfrequency, quality factor, as well as the vibrational amplitude. The eigenfrequency response to a changing effective spring constant is most interesting and is discussed in the section.

There are generally two mechanisms that produce a change of the effective spring constant of a nanomechanical resonator. First, the effective spring constant can be altered by force gradients. Second, the effective spring constant of mechanical resonators is sensitive to changes of temperature T. On the one hand, temperature influences the material's Young's modulus. On the other hand, changes of temperature induce thermoelastic responses in the material. Both of these mechanisms are discussed below.

6.2.1 *Response to Force Gradient ($\mathscr{R}_{F'}$)*

The detection of a force gradient is best known from dynamic atomic force microscopy, where the cantilever-tip interaction with the substrate produces a shift of the cantilever resonance frequency [17]. Assuming a small mechanical vibration z around the equilibrium position of the resonator at z_0, the static force to be measured can be approximated by a first-order Taylor series:

$$F(z_0 + z) \approx F(z_0) + \frac{\partial F(z_0)}{\partial z} z. \tag{6.33}$$

Apparently, if the force has a nonzero gradient component $\partial F(z_0)/\partial z = F'$, then the equation of motion of a lumped-element resonator (1.1) becomes:

$$m\ddot{z} + c\dot{z} + kz = F(z_0) + F' z$$
$$m\ddot{z} + c\dot{z} + (k - F') z \approx 0.$$
(6.34)

The force gradient is altering the effective spring constant of the mechanical resonator. On the right side of the equation remains a static force $F(z_0)$. This force pushes the mechanical resonator into a new equilibrium point. Assuming that this force is small, the induced static deflection is negligible for a linear system. The eigenfrequency of the system can then be obtained from the homogenous differential equation (6.34), which readily results in the eigenfrequency as a function of the force gradient (see Sect. 1.1.1 on page 3 for more detail):

$$\omega_0(F') = \sqrt{\frac{k - F'}{m}} = \omega_0(0)\sqrt{1 - \frac{F'}{k}}.$$
(6.35)

Assuming that $F'/k \ll 1$, the eigenfrequency can be approximated by the first-order Taylor series:

$$\omega_0(F') \approx \omega_0(0)\left(1 - \frac{F'}{2k}\right).$$
(6.36)

This simple derivation predicts that a force gradient $F' = \partial F(z_0)/\partial z$ causes a detuning of the eigenfrequency to a lower value. Force gradients can occur, as mentioned in the beginning, e.g., at the contact point between a surface and a atomic force microscopy tip. These gradients are often of electrostatic nature. Hence, a similar effect occurs with electrostatic transduction, as discussed in the next subsection.

Finally, the responsivity can readily be calculated from (6.36)

$$\mathcal{R}_{F'} = \frac{\partial \omega_0(F')}{\partial F'}\frac{1}{\omega_0(0)} = -\frac{1}{2k},$$
(6.37)

and a change in the force gradient can be detected by measuring the fractional frequency change y

$$\Delta F' = y\,\mathcal{R}_{F'}^{-1}.$$
(6.38)

6.2.1.1   Electrostatic Force Gradients in Parallel Plate Capacitors

Electrostatic forces are common transduction schemes to actuate micro- and nanomechanical resonators, as it is discussed in more detail in Sect. 4.2.1.1 on page 113. It is possible to tune the resonance frequency by applying a d.c. bias voltage tuning [18, 19]. The same effect can also be used to measure the electrostatic force strength as a function of the applied bias potential [20]. In this subsection, the

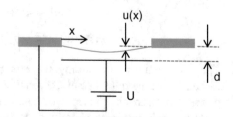

Fig. 6.8 Schematic drawing of a flexural resonator placed in a distance d above a conducting plate. There is an electrostatic potential V applied between the resonator and the plate

frequency response of a membrane resonator to a d.c. bias voltage is derived, similar to the experiment performed in [20, 21].

Neglecting any electrostatic fringe fields, the electrostatic force between two parallel plates under a potential difference V, of area A, and separated by a distance d is given by (4.15)

$$F(U) = -\frac{1}{2}\varepsilon_0\varepsilon_r \frac{A}{(d + U(x, y, t))^2} V^2, \tag{6.39}$$

with the permittivity of vacuum ε_0 and the relative permittivity of the surrounding medium ε_r.

A schematic of the membrane resonator is shown in Fig. 6.8. The equilibrium of forces for an infinitesimal piece of membrane with the area $dx \times dy$ and thickness h is then given by

$$\rho h \frac{\partial^2 U}{\partial t^2}(x, y, t) - \sigma h \nabla^2 U(x, y, t) = F(U(x, y, t)) \tag{6.40}$$

with the displacement function $U(x, y, t)$, the tensile pre-stress σ, and the mass density ρ. The deflection of a membrane can be described by (2.70)

$$U(x, y, t) = \sum_{n=1}^{\infty} \sum_{j=1}^{\infty} u_{0,n,j} \, \phi_{n,j}(x, y) \, e^{i\omega_{n,j}t} \tag{6.41}$$

with the normalized modeshape function

$$\phi_{n,j}(x, y) = \sin\frac{n\pi x}{L_x} \sin\frac{j\pi y}{L_y}. \tag{6.42}$$

Considering the first-order Taylor approximation of the electrostatic force (6.39), the equation of motion can be written as

$$\rho h \frac{\partial^2 U}{\partial t^2} - \sigma h \nabla^2 U = F(0) + \frac{\partial F(0)}{\partial U} U$$

$$\rho h \frac{\partial^2 U}{\partial t^2} - \sigma h \nabla^2 U = -\frac{\varepsilon_0\varepsilon_r}{2d^2} V^2 + \frac{\varepsilon_0\varepsilon_r}{d^3} V^2 U \tag{6.43}$$

In a "linear system," the static force term $-\frac{\varepsilon_0\varepsilon_r}{2d^2}V^2$ causes a static deflection of the membrane. This static deflection does not influence the eigenfrequency and can thus be neglected. Following Galerkin's method, as introduced in Sect. 2.6.2 on page 61, (6.40) can be solved for the fundamental normal mode by multiplying it with $\phi_{1,1}$ and integrating over the entire membrane area $A = L \times L$ with $L = L_x = L_y$. Taking the scalar product of the fundamental mode gives:

$$\iint_A \left(\rho h \frac{\partial^2 U}{\partial t^2} - \sigma h \nabla^2 U + \frac{\varepsilon_0\varepsilon_r}{d^3}V^2 U \right) \phi_{1,1} dx\, dy = 0 \qquad (6.44)$$

which with (6.41) can be written as

$$\rho h \omega_{1,1}^2 \iint_A \phi_{1,1}^2 dx\, dy - 2\frac{\pi^2}{L^2}\sigma h \iint_A \phi_{1,1}^2 dx\, dy + \frac{\varepsilon_0\varepsilon_r}{d^3}V^2 \iint_A \phi_{1,1}^2 dx\, dy = 0$$

$$\rho h \omega_{1,1}^2 - 2\frac{\pi^2}{L^2}\sigma h + \frac{\varepsilon_0\varepsilon_r}{d^3}V^2 = 0. \qquad (6.45)$$

The eigenfrequency can now readily be isolated from (6.45)

$$\omega_{1,1}^2(V) = 2\pi^2 \frac{\sigma}{\rho}\frac{1}{L^2} - \frac{\varepsilon_0\varepsilon_r}{\rho h d^3}V^2 \qquad (6.46)$$

With the eigenfrequency of a membrane (2.87) with zero voltage applied

$$\omega_{1,1}(0) = \frac{\sqrt{2}\pi}{L}\sqrt{\frac{\sigma}{\rho}} \qquad (6.47)$$

the first-order Taylor approximation of $\omega_{1,1}(V)$ becomes:

$$\omega_{1,1}(V) \approx \omega_{1,1}(0)\left(1 - \frac{\varepsilon_0\varepsilon_r L^2}{4\pi^2 h d^3 \sigma}V^2 \right). \qquad (6.48)$$

From (6.48) it can be seen that an applied bias voltage V causes a negative frequency detuning.

6.2.2 Response to Temperature ($\mathscr{R}_T$)

As the quality factor of micro- and nanomechanical resonators is a function of temperature, as introduced in Sect. 3.3 on page 81, so is the resonance frequency. The temperature responsivity of the resonance frequency is caused by two main effects: (i) the Young's modulus as a function of temperature $E(T)$, typically a material softens with increasing temperature, and (ii) a temperature change ΔT induces a strain from the thermal expansion of the material $\varepsilon = \alpha\,\Delta T$. Depending on the configuration, a thermal expansion results in a change of geometry or stress. For most applications a high-frequency stability is required, and hence resonators with a small temperature response are sought after. A high temperature responsivity can impede a particular application if a small ambient temperature change, e.g., significantly changes the frequency of a filter or produces a fake sensor signal. Typically, sensors are temperature stabilized either by (i) using a temperature-insensitive resonator design, (ii) performing a differential measurement between two resonators that both are exposed to the same temperature bath but only one of them is sensing, (iii) the measurement signal is temperature corrected by means of an integrated temperature sensor, or (iv) the ambient temperature directly is controlled and stabilized. Besides a change in ambient temperature, a significant local heating can also be introduced by the electric or optic transduction of the nanomechanical resonator.

In contrast to the sensor applications mentioned above, a high as possible temperature responsivity is the figure of merit if the resonator is actually used to detect temperature changes. This is the case if micro- and nanomechanical resonators are used as thermometer [22, 23] or to measure local heating, as will be discussed in the next Sect. 6.2.3.

In all cases and for all applications, the understanding of the temperature response of different resonators is crucial. Here, the temperature responsivity to distinct cases, unstressed and stressed nanomechanical resonators, is introduced.

6.2.2.1   Response to Temperature of Stressed Resonators

Strings and drumheads show a particularly strong response to temperature change. The thermal expansion is directly changing the tensile stress, which is the core parameter defining a string's or drumhead's resonance frequency.

The eigenfrequency of a string is given by (2.55), and it is mainly defined by the pre-stress σ. Different thermal expansion of the beam and the supporting chip make the strain of the beam temperature-dependent which will result in a temperature-dependent tensile stress. A schematic of such a string is shown in Fig. 6.9. The eigenfrequency will therefore also be a function of temperature. If the beam and the frame expand linearly with temperature, the temperature-dependent strain, assuming the string and frame are in thermal equilibrium, is given as

$$\varepsilon(T) = \varepsilon_0 - (\alpha_{str} - \alpha_{fr})(T - T_0) \tag{6.49}$$

where ε_0 is the strain at temperature T_0, T the temperature and α_i the coefficient of thermal expansion of the beam and the support. Stress and strain are related to each other via Hooke's laws if the beam is made of a linear elastic material and the strain does not exceed the elastic limits. If this is true, the temperature-dependent stress can be defined as

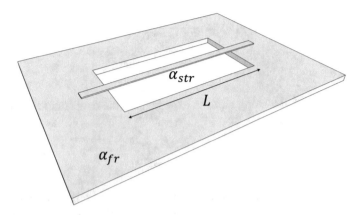

Fig. 6.9 Schematic drawing of a string with length L and a thermal expansion coefficient α_{str} spanned by a frame with a thermal expansion coefficient α_{fr}

$$\sigma(T) = E\varepsilon(T)$$
$$\sigma(T) = E(\varepsilon_0 - (\alpha_{str} - \alpha_{fr})(T - T_0))$$
$$\sigma(T) = \sigma_0 - E(\alpha_{str} - \alpha_{fr})(T - T_0). \tag{6.50}$$

Inserting the temperature-dependent stress into (2.55), the temperature-dependent eigenfrequency is given by

$$\omega_n(T) = \frac{n\pi}{L}\sqrt{\frac{\sigma_0 - E(\alpha_{str} - \alpha_{fr})(T - T_0)}{\rho}}, \tag{6.51}$$

which for a small stress change, which is for a small ΔT, can be approximated by the first-order Taylor approximation:

$$\omega_n(T) \approx \omega_n(T_0)\left(1 - \frac{1}{2}\frac{E(\alpha_{str} - \alpha_{fr})(T - T_0)}{\sigma_0}\right) \tag{6.52}$$

In this approximation temperature-induced changes in length, Young's modulus, coefficients of thermal expansion, and density have been neglected. For a real beam, it can be expected that the length and the coefficients of thermal expansion will be increasing with temperature and that Young's modulus and the density will be decreasing with temperature. Neglecting these changes will introduce a small error when estimating the resonance frequency.

The temperature responsivity of a tensile stressed beam and string can be written as

$$\mathscr{R}_T(\omega) = \frac{\partial \omega_n}{\partial T} \frac{1}{\omega_n(T_0)}$$

$$= -\frac{1}{2}(\alpha_{str} - \alpha_{fr})\frac{E}{\sigma_0}. \tag{6.53}$$

For the case that the string alone is experiencing the temperature change, with $\alpha_{str} = \alpha$, (6.53) reduces to

$$\mathscr{R}_T(\omega) = -\frac{\alpha E}{2\sigma_0}. \tag{6.54}$$

From this expression it can be seen that the temperature responsivity of a string is increasing with the Young's modulus and a difference in thermal expansion and for a decreasing pre-stress. In fact, as the ratio of $E/\sigma_0 \gg 1$, the responsivity in strings (6.54) is significantly enhanced over unstressed resonators (6.59). The responsivity is negative if the string has a higher coefficient of thermal expansion than the frame and vice versa. A temperature change can then be measured with a string resonator by

$$\Delta T = y\, \mathscr{R}_T^{-1}, \tag{6.55}$$

for a fractional frequency change y. Temperature sensitivities in the μK range have been demonstrated with silicon nitride string resonators [23].

The temperature responsivity for a drumhead resonator with biaxial tensile stress, for the situation that only the drumhead itself experience a temperature change while the frame temperature remains constant, is given by [25–27]

$$\mathscr{R}_T(\omega) = -\frac{\alpha E}{2\sigma_0(1 - \nu)}, \tag{6.56}$$

where ν is the Poisson ratio.

These models can readily be extended for resonators that are made of multiple layers of different materials by choosing effective material parameters [4] .

6.2.2.2 Response to Temperature of Unstressed Resonators

Stress-released structures, such as singly clamped beams (cantilevers), as introduced in Sect. 2.2, are among the least temperature-sensitive resonators. In the first order of approximation, for any cantilever the temperature-induced strain can relax, which only results in a slight change of geometry. In this case, it is possible to calculate the eigenfrequency (2.31) after the thermal expansion of the material after a change of temperature $\Delta T = T - T_0$:

$$\omega_n(T) = \lambda_n^2 \sqrt{\frac{E(T)\,I_y(1+\alpha\Delta T)^4}{\rho(1+\alpha\Delta T)^{-3}A(1+\alpha\Delta T)^2 L^4(1+\alpha\Delta T)^4}}$$

$$= \lambda_n^2 \sqrt{\frac{E(T)I_y}{\rho A L^4}}\sqrt{(1+\alpha\Delta T)} \tag{6.57}$$

with the temperature-sensitive Young's modulus with an assumed linear thermal softening coefficient α_E

$$E(T) = E_0(1 + \alpha_E \Delta T). \tag{6.58}$$

The temperature responsivity is then given by

$$\mathscr{R}_T(\omega) = \left.\frac{\partial\omega_n(T)}{\partial T}\right|_{T=T_0} \frac{1}{\omega_n(T_0)} \approx \frac{\alpha+\alpha_E}{2}. \tag{6.59}$$

Both effects are equally weighted and act on the eigenfrequency proportional to the square root of the specific thermal coefficient. However, as said coefficients are very small, the Taylor expansion can be truncated at the first order. Typically, the temperature-induced change in the Young's modulus is dominating over the geometrical effect from the thermal expansion. For example, for silicon, the thermal expansion coefficient is $\alpha = 2.6\,\text{ppm/K}$ and the thermal softening coefficient is $\alpha_E \approx -44\,\text{ppm/K}$[28]. Hence, the temperature effect on the Young's modulus is roughly one order of magnitude larger than the effect of thermal expansion. The same is true, e.g., for silicon nitride with $\alpha = 2.3\,\text{ppm/K}$ and $\alpha_E \approx -87\,\text{ppm/K}$[29]. The two mechanisms involved are acting in the opposite way. On the one hand, the thermal expansion is causing a frequency increase with increasing temperature, while on the other hand the softening of the Young's modulus is lowering the frequency with increasing temperature.

The analysis, as it is here performed for cantilevers, holds true for all stress released structures, e.g., one-dimensional bulk vibrations (Sect. 2.4), torsional beam vibrations (Sect. 2.5), or bending plate vibrations (Sect. 2.3).

In some cases, however, it is necessary to look into more detail due to the fact that the stress is never completely released at the clamping. This indeed gives rise to a dependence of the frequency on the stress that might be dominant over the previously analyzed dependencies. In addition, often micro- and nanomechanical structures aren't uniform, but are made from multiple layers of materials with different thermal expansion coefficients. A common design consists, e.g., of dielectric structural materials, such as SiN and SiC, which is coated with a thin metal layer. In this case, the modeling of the temperature response becomes more difficult, as a temperature change causes a complex stress field and resulting deformation with a hard to predict influence on the eigenfrequency. This effect has, e.g., been observed in 50- nm-thick silicon nitride cantilevers coated with a 20- nm-thick gold layer. The local heating with a readout laser resulted in a substantial detuning of the resonance frequency [30].

Finally, in the case of doubly clamped beams (bridges), the analysis also holds true for temperature changes that cause only a small stress, where small here is defined as

$$\sigma \ll \frac{E I_y n^2 \pi^2}{A L^2} \approx \frac{E n^2 \pi^2}{12} \left(\frac{t}{L} \right)^2$$
$$\Delta T \ll \frac{I_y n^2 \pi^2}{\alpha A L^2} \approx \frac{n^2 \pi^2}{12\alpha} \left(\frac{t}{L} \right)^2$$

(6.60)

as can be extracted from (2.51), and by assuming a linear elastic material with a linear thermal expansion $\sigma = E\alpha \Delta T$. As an example, this gives for a silicon bridge with a ratio of length to height of $L/h = 100$ a pretty robust temperature range of roughly $\Delta T \ll 50\,\text{K}$. But as soon as the length to height ratio becomes larger, the temperature-induced stress starts to significantly contribute to the eigenfrequency of the resonator, and hence a small temperature change results in a large change of the frequency. Strings and membranes are the extreme cases where the tensile stress is the dominating parameter that defines the eigenfrequency. Hence, strings and membranes are highly responsive to changes of stress and hence to changes of temperature. This case will be discussed in more detail in the following subsection.

6.2.3 Response to Local Heating ($\mathscr{R}_P$)

In this section, the effect of local heating on the resonance frequency of a nanomechanical resonator is discussed. Local heating is the insertion of thermal energy at a specific location on the resonator. Heating can be induced, e.g., by resistive heating [31] or by the absorption of radiation, which can be an unwanted or deliberate effect.

The former situation plays an important role in many experiments where the resonator vibration is detected optically, e.g., with an integrated optical ring resonator or a free-space laser, as depicted in the schematics in Fig. 6.10a–c. The former configuration is often used in optomechanics [32–34]. Here fluctuations in the laser power can result in frequency noise of the mechanical resonator [35]. In the latter situation, a high responsivity to local heating is desired for sensing applications, such as thermal detection of electromagnetic radiation (as schematically depicted in Fig. 6.10d) [25, 36, 37] or for photothermal absorption microscopy and spectroscopy [38–42]. Another scenario in which a high temperature response is wanted is for thermal frequency tuning [43].

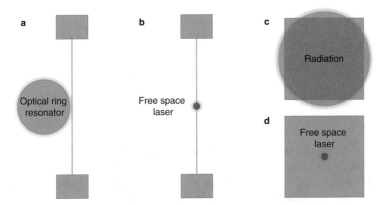

Fig. 6.10 Different scenarios of local heating of nanomechanical resonators. (**a**) Local heating of optomechanical string resonator due to evanescent coupling to optical ring resonator. (**b**) Local heating of string resonator by free-space laser. (**c**) Local heating of drumhead resonator by free-space laser. (**d**) Heating of drumhead resonator, e.g., by infrared radiation

The frequency response of a nanomechanical resonator to heating is directly related to the response to a change in temperature, which was discussed in the previous Sect. 6.2.2. The finite heat capacity filters all temperature variations above the thermal frequency $1/\tau_{th}$. The responsivity to thermal energy insertion with power P is then given by [44, 45]

$$\mathscr{R}_P(\omega) = \frac{\partial \omega_0}{\partial P} \frac{1}{\omega_0} H_{th}(\omega) = \frac{\partial \omega_0}{\partial T} \frac{1}{\omega_0} \frac{\partial T}{\partial P} H_{th}(\omega) = \frac{\mathscr{R}_T(\omega)}{G} H_{th}(\omega). \qquad (6.61)$$

The composition of the thermal conductance depends on the experimental circumstances. As most nanomechanical resonators are operated in vacuum, G typically is the sum of heat transfer by thermal conductance and thermal radiation. Especially for structures with a large aspect ratio, such as strings and drumhead resonators, thermal radiation can become the dominating heat transfer mechanism [45, 46].

In case the thermal energy is coming from absorption of electromagnetic radiation, the responsivity depends on the emissivity ε of the nanomechanical resonators at the radiation-specific wavelength λ

$$\mathscr{R}_{P_\varepsilon}(\omega) = \varepsilon \mathscr{R}_P(\omega). \qquad (6.62)$$

According to Kirchhoff's law, the emissivity is equal to the absorptivity for the case that the nanomechanical resonator is in thermodynamic equilibrium with its surroundings.

Assuming the special case of dominating temperature fluctuation noise (5.57 on page 172), the sensitivity (6.6), with (6.61) and (6.62), can be expressed as

$$S_P(\omega) = \frac{S_{y_{thf}}(\omega)}{\left[\mathscr{R}_{P_\varepsilon}(\omega)\right]^2}$$

$$= S_T \left(\frac{\mathscr{R}_T(\omega)}{\mathscr{R}_{P_\varepsilon}(\omega)}\right)^2 \tag{6.63}$$

$$= \frac{4k_B T^2 G}{\varepsilon^2}.$$

This is the sensitivity of a thermal detector, whose square root is equivalent to the noise equivalent power NEP [24]. In case the resonator is well thermally isolated and hence heat transfer is dominated by thermal radiation, the thermal conductance is given by the derivative of the Stefan-Boltzmann law:

$$G_{rad} = \frac{d}{dT} A_s \varepsilon \sigma_B T^4 = 4 A_s \varepsilon \sigma_B T^3 \tag{6.64}$$

where $\sigma_B = 5.67 \times 10^{-8}$ W/(m^2 K^4) is the Stefan-Boltzmann constant. With the heat conductance $G = G_{rad}$ and (6.63), the noise equivalent power becomes [24, 47]:

$$NEP = \sqrt{S_P} = \left(\frac{16\sigma_B k_B A_s T^5}{\varepsilon}\right)^{1/2}, \tag{6.65}$$

which when expressed as the *specific detectivity*

$$D^* = \frac{\sqrt{A_s}}{NEP} = \left(\frac{\varepsilon}{16\sigma_B k_B T^5}\right)^{1/2}. \tag{6.66}$$

The specific detectivity is the figure of merit of photon detectors, whose NEP scales with the detector size. D^* for quantum-limited detectors is independent of the detector area A_s. Because of the quantum nature of the thermal fluctuation noise, the specific detectivity (6.66) of a thermal detector becomes also independent of the lateral size. This is not the case if the noise is not limited by quantum noise. Equation (6.66) is the specific detectivity of a thermal detector that is coupled to the environment via a single side, which yields the famous ultimate sensitivity limit of thermal detectors at room temperature ($T = 290$ K and $\varepsilon = 1$) of $D^* \approx 2 \times 10^{10}$ cm Hz$^{1/2}$/W [24, 47].

6.2.3.1 Local Heating at String Center

As mentioned in the introduction to this section, the heating of a string at the center is a common scenario that occurs when an electromagnetic energy is absorbed, e.g., from a readout laser or during a photothermal analysis experiment, as schematically depicted in Fig. 6.10a and b.

In vacuum, heat transfer by convection can be omitted. The heat transfer mechanisms remaining that have to be considered are thermal radiation and thermal conduction with the partial thermal conductances G_{rad} and G_{cond}, respectively.

Let's first discuss heat transfer by conduction through the strings. Assuming a stationary situation in which the energy P, absorbed in the string center, is in equilibrium with the heat flow through the strings out into the frame of constant temperature T_0. In this case the temperature field in the string can be readily described by a linear function:

$$T(x) = T_0 + 2(T_1 - T_0)\frac{x}{L} = T_0 + 2\Delta T \frac{x}{L} \quad \text{for} \quad 0 \le x \le \frac{L}{2} \tag{6.67}$$

as it is depicted in Fig. 6.11.

The resulting heat flux for the central temperature increase ΔT through the two halves of length $(L/2)$ of the string with a cross section $A = hw$ is described by one-dimensional Fourier's law. For the given temperature field (6.67), this yields the conductive heat flow:

$$G_{cond} = 4\kappa \frac{wh}{L} \tag{6.68}$$

with the thermal conductivity κ of the string.

The radiative conductance for the specific temperature field (6.67) can be derived from (6.64). On the one hand, the temperature of the entire string effectively increases by $\langle T \rangle = \Delta T/2$. On the other hand, the heat radiates from both upper and lower surfaces of the string, whereby the radiation from the thin side surfaces are neglected. The radiative conductance can now be written as

$$G_{rad} = 4\sigma_B \varepsilon T^3 Lw. \tag{6.69}$$

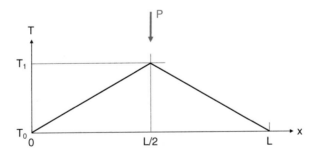

Fig. 6.11 Temperature distribution in a string of length L with a point heat source of power P in the center

Finally, the total thermal conductance, with respect to the temperature increase ΔT in the string center, is given by

$$G_{\Delta T} = G_{cond} + G_{rad} = 4w \left[\kappa \frac{h}{L} + \sigma_B \varepsilon_{rad} T^3 L \right]. \tag{6.70}$$

This readily yields the temperature increase ΔT in the center of the string as a function of the absorbed power P:

$$\Delta T(P) = \frac{P}{G_{\Delta t}} = \frac{P}{4w} \left[\kappa \frac{h}{L} + \sigma_B \varepsilon_{rad} T_0^3 L \right]^{-1}. \tag{6.71}$$

Because the string does not experience an even temperature increase, the responsivity to temperature change has to be adjusted with a factor of $1/2$ to correct for the effective string temperature $\langle T \rangle = \Delta T/2$. With that, combining (6.70) and the adjusted responsivity to temperature (6.54), the responsivity to local heating (6.61) can be calculated:

$$\mathscr{R}_P(\omega) = \frac{\mathscr{R}_T(\omega)}{2G} H_{th}(\omega) = -\frac{1}{16} \frac{\alpha E}{\sigma w} \left[\kappa \frac{h}{L} + \sigma_B \varepsilon_{rad} T_0^3 L \right]^{-1} H_{th}(\omega). \tag{6.72}$$

This formula is plotted in Fig. 6.12 for a nanomechanical silicon nitride string. This example shows that the responsivity to an absorbed power in the string center has two regimes. For short strings, heat transfer by thermal conduction dominates and the fractional responsivity increases linearly with string length. For long strings, the thermal conductivity becomes small enough such that heat transfer through thermal radiation becomes dominant. Generally, (6.72) predicts that the responsivity to an

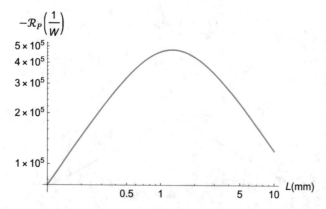

Fig. 6.12 Fractional responsivity of a string for absorbed power in its center (6.72) as a function of its length, assuming a low-stress ($\sigma = 200\,\text{MPa}$) silicon nitride string that is 50 nm thick and 1 µm wide with the following parameters $\alpha = 1.2 \times 10^{-6}\,\text{K}^{-1}$, $\kappa = 2.5\,\text{W}\,\text{m}^{-1}\,\text{K}^{-1}$, $\varepsilon_{rad} = 0.05$ [46], $E = 250\,\text{GPa}$, $T_0 = 300\,\text{K}$

absorbed power in the string center is maximal for narrow strings with a low tensile stress σ and a large thermal expansion coefficient α.

6.2.3.2 Even Heating of Drumhead

The scenario of a drumhead that is evenly heated, e.g., through radiation, is shown in Fig. 6.10c. An even heating can either be deliberate, e.g., for detection of electromagnetic radiation [25, 37, 48], or as inevitable thermal radiation from the surrounding [26, 45, 46].

In contrast to the local point heating discussed in the previous section, for an even heating, the heat capacity is given by the total drumhead's thermal mass [45]:

$$C_{th} = \rho c_p L^2 h. \tag{6.73}$$

The thermal response time of a drumhead has been derived considering conductive and radiative heat transfer [46]:

$$\tau_{th} = \left(\underbrace{\frac{2\pi^2}{L^2} \frac{\sum_i h_i \kappa_i}{\sum_i h_i \rho_i C_{p_i}}}_{\text{Conduction}} + \underbrace{\frac{8\epsilon\sigma_B T^3}{\sum_i h_i \rho_i c_{p_i}}}_{\text{Radiation}} \right)^{-1}, \tag{6.74}$$

with the Stefan-Boltzman constant σ_B, the emissivity of the drumhead surface ε, the mass density ρ, the specific heat capacity c_p for each layer i and the temperature T.

Based on this heat capacity and the thermal time constant (6.74), the thermal conductance can be calculated from

$$G = \frac{C_{th}}{\tau_{th}} = 2\pi^2 \kappa h + 8\varepsilon L^2 \sigma_B T^3. \tag{6.75}$$

Finally, with the thermal conductance, the responsivity can be calculated from (6.61)

$$\begin{aligned} \mathcal{R}_P(\omega) &= \frac{\mathcal{R}_T(\omega)}{G} H_{th}(\omega) \\ &= \frac{\alpha E}{4\sigma(1-\nu)} \left[\pi^2 \kappa h + 4\varepsilon L^2 \sigma_B T^3 \right]^{-1} H_{th}(\omega). \end{aligned} \tag{6.76}$$

6.2.3.3 Local Heating of Drumhead Center

The local heating of a rectangular drumhead is depicted in Fig. 6.10d. The thermomechanical behavior of a square drumhead can be modeled as a circular drumhead as good approximation [49]. Assuming heat transfer by thermal conduction alone, the temperature field of a circular drumhead of radius $a = L/2$ as a function of the absorbed power P for the distance from the center r is given by [50]

$$T(r) = -\frac{P}{2\pi\kappa h}\ln\frac{r}{a}.$$ (6.77)

This equation is valid for $r > b$, where b is the radius of the absorbed heating source, e.g., a focused laser beam.

Based on the temperature field (6.77) and assuming that heat transfer is dominated by thermal conduction, the responsivity of such a circular drumhead has been derived analytically, and is given by [50]

$$\mathcal{R}_{P_{cond}}(\omega) = -\frac{1}{8\pi}\frac{\alpha E}{\kappa h\sigma}\left(\frac{1.36 - 0.36\nu}{1-\nu}\right)H_{th}(\omega).$$ (6.78)

It is noteworthy that the fractional responsivity of a drumhead is independent of the lateral size.

With the thermal time constant (6.74) for a single material and considering conductive heat transfer alone

$$\tau_{th_{cond}} = \frac{L^2\rho c_p}{2\pi^2\kappa},$$ (6.79)

the responsivity for absorbed power (6.61) can be estimated as follows:

$$\begin{aligned}
\mathcal{R}_P(\omega) &= \frac{\mathcal{R}_T(\omega)}{G}H_{th}(\omega)\\
&= \mathcal{R}_{P_{cond}}(\omega)\frac{\tau_{th}}{\tau_{th_{cond}}}\\
&= \frac{(1.36 - 0.36\nu)\pi}{4}\frac{\alpha E}{2\sigma(1-\nu)}\left[\pi^2\kappa h + 4\varepsilon L^2\sigma_B T^3\right]^{-1}H_{th}(\omega)\\
&\approx \frac{\alpha E}{2\sigma(1-\nu)}\left[\pi^2\kappa h + 4\varepsilon L^2\sigma_B T^3\right]^{-1}H_{th}(\omega).
\end{aligned}$$ (6.80)

It shows that the responsivity to a focused heating source in the center of a drumhead is approximately twice that of an even heating (6.76) for the same power P.

References

1. Moser, J. Güttinger, J., Eichler, A., Esplandiu, M. J., Liu, D. E., Dykman, M. I., & Bachtold, A. (2013). Ultrasensitive force detection with a nanotube mechanical resonator. *Nature Nanotechnology, 8*(7), 493–496.

2. Héritier, M., Eichler, A., Pan, Y., Grob, U., Shorubalko, I., Krass, M. D., Tao, Y., & Degen, C. L. (2018). Nanoladder cantilevers made from diamond and silicon. *Nano Letters, 18*(3), 1814–1818.
3. Schmid, S., & Hierold, C. (2008). Damping mechanisms of single-clamped and prestressed double-clamped resonant polymer microbeams. *Journal of Applied Physics, 104*(9), 093516.
4. Bose, S., Schmid, S., Larsen, T., Keller, S. S., Sommer-Larsen, P., Boisen, A., & Almdal, K. (2014). Micromechanical string resonators: Analytical tool for thermal characterization of polymers. *ACS Macro Letters, 3*(1), 55–58.
5. Tao, Y., Navaretti, P., Hauert, R., Grob, U., Poggio, M., & Degen, C. L. (2015). Permanent reduction of dissipation in nanomechanical Si resonators by chemical surface protection. *Nanotechnology, 26*(46), 465501.
6. Hanay, M. S., Kelber, S., Naik, A. K., Chi, D., Hentz, S., Bullard, E. C., Colinet, E., Duraffourg, L., & Roukes, M. L. (2012). Single-protein nanomechanical mass spectrometry in real time. *Nature Nanotechnology, 7*, 602–608.
7. Schmid, S., Kurek, M., Adolphsen, J. Q., Boisen, A. (2013). Real-time single airborne nanoparticle detection with nanomechanical resonant filter-fiber. *Scientific Reports, 3*, 1288.
8. Dohn, S., Sandberg, R., Svendsen, W., & Boisen, A. (2005). Enhanced functionality of cantilever based mass sensors using higher modes. *Applied Physics Letters, 86*(23), 233501.
9. Dohn, S., Svendsen, W., Boisen, A., & Hansen, O. (2007). Mass and position determination of attached particles on cantilever based mass sensors. *Review of Scientific Instruments, 78*(10), 103303.
10. Yu, H., & Li, X. (2009). Bianalyte mass detection with a single resonant microcantilever. *Applied Physics Letters, 94*(1), 173–176.
11. Schmid, S., Dohn, S., & Boisen, A. (2010). Real-time particle mass spectrometry based on resonant micro strings. *Sensors, 10*(9), 8092–8100.
12. Dohn, S., Schmid, S., Amiot, F., & Boisen, A. (2010). Position and mass determination of multiple particles using cantilever based mass sensors. *Applied Physics Letters, 97*(4), 044103.
13. Hanay, M. S., Kelber, S. I., O'Connell, C. D., Mulvaney, P., Sader, J. E., & Roukes, M. L. (2015). Inertial imaging with nanomechanical systems. *Nature Nanotechnology, 10*(4), 339–344.
14. Schmid, S., Kurek, M., & Boisen, A. (2013). Towards airborne nanoparticle mass spectrometry with nanomechanical string resonators. *SPIE Defense, Security, and Sensing, 8725*, 872525–872528.
15. Yang, Y. T., Callegari, C., Feng, X. L., Ekinci, K. L., & Roukes, M. L. (2006). Zeptogram-scale nanomechanical mass sensing. *Nano Letters, 6*(4), 583–586.
16. Li, M., Tang, H. X., & Roukes, M. L. (2007). Ultra-sensitive NEMS-based cantilevers for sensing, scanned probe and very high-frequency applications. *Nature Nanotechnology, 2*(2), 114–120.
17. Voigtlaender, B. (2015). *Scanning probe microscopy*. Berlin: Springer.
18. Kozinsky, I., Postma, H. W. Ch., Bargatin, I., & Roukes, M. L. (2006). Tuning nonlinearity, dynamic range, and frequency of nanomechanical resonators. *Applied Physics Letters, 88*(25), 253101.
19. Unterreithmeier, Q. P., Weig, E. M., & Kotthaus, J. P. (2009). Universal transduction scheme for nanomechanical systems based on dielectric forces. *Nature, 458*(7241), 1001–1004.
20. Schmid, S., Bagci, T., Zeuthen, E., Taylor, J. M., Herring, P. K. Cassidy, M. C., Marcus, C. M., Guillermo Villanueva, L., Amato, B., Boisen, A., Cheol Shin, Y., Kong, J., Sørensen, A. S., Usami, K., & Polzik, E. S. (2014). Single-layer graphene on silicon nitride micromembrane resonators. *Journal of Applied Physics, 115*(5), 054513.
21. Bagci, T., Simonsen, A., Schmid, S., Villanueva, L. G., Zeuthen, E., Appel, J., Taylor, J. M., Sørensen, A., Usami, K., Schliesser, A., & Polzik, E. S. (2014). Optical detection of radio waves through a nanomechanical transducer. *Nature, 507*(7490), 81–85.
22. Pandey, A. K., Gottlieb, O., Shtempluck, O., & Buks, E. (2010). Performance of an AuPd micromechanical resonator as a temperature sensor. *Applied Physics Letters, 96*, 203105.

23. Larsen, T., Schmid, S., Gronberg, L., Niskanen, A. O., Hassel, J., Dohn, S., & Boisen, A. (2011). Ultrasensitive string-based temperature sensors. *Applied Physics Letters, 98*, 121901.
24. Kruse, P. W., McGlauchlin, L. D., & McQuistan, R. B. (1962). *Elements of infrared technology: Generation, transmission and detection.* New York: Wiley.
25. Blaikie, A., Miller, D., & Alemán, B. J. (2019). A fast and sensitive room-temperature graphene nanomechanical bolometer. *Nature Communications, 10*(1), 1–8.
26. Zhang, C., Giroux, M., Nour, T. A., & St-Gelais, R. (2020). Radiative heat transfer in freestanding silicon nitride membranes. *Physical Review Applied, 14*(2), 024072.
27. Ventsel, E., & Krauthammer, T. (2001). *Thin plates and shells : Theory, analysis, and applications.* New York : Marcel Dekker.
28. Rendón, M., & Makarov, N. (2014). Determination of the Si Young's modulus between room and melt temperature using the impulse excitation technique. *Physica Status Solidi (C), 11*(1), 150–155.
29. Chuang, W. -H., Luger, T., Fettig, R. K., & Ghodssi, R. (2004). Mechanical property characterization of LPCVD silicon nitride thin films at cryogenic temperatures. *Journal of Microelectromechanical Systems, 13*(5), 870–879.
30. Pini, V., Tamayo, J., Gil-Santos, E., Ramos, D., Kosaka, P., Tong, H. -D., van Rijn, C., & Calleja, M. (2011). Shedding light on axial stress effect on resonance frequencies of nanocantilevers. *ACS Nano, 5*(6), 4269–4275.
31. St-Gelais, R., Bernard, S., Reinhardt, C., & Sankey, J. C. (2019). Swept-frequency drumhead optomechanical resonators. *ACS Photonics, 6*(2), 525–530.
32. Brawley, G. A., Vanner, M. R., Larsen, P. E., Schmid, S., Boisen, A., & Bowen, W. P. (2014). Non-linear optomechanical measurement of mechanical motion. arXiv:1404.5746.
33. Zhang, R., Ti, C., Davanço, M. I., Ren, Y., Aksyuk, V., Liu, Y., & Srinivasan, K. (2015). Integrated tuning fork nanocavity optomechanical transducers with high fMQM product and stress-engineered frequency tuning. *Applied Physics Letters, 107*(13), 131110.
34. Gavartin, E., Verlot, P., & Kippenberg, T. J. (2012). A hybrid on-chip optomechanical transducer for ultrasensitive force measurements. *Nature Nanotechnology, 7*, 509–514.
35. Sadeghi, P., Demir, A., Guillermo Villanueva, L., Kähler, H., & Schmid, S. (2020). Frequency fluctuations in nanomechanical silicon nitride string resonators. *Physical Review B, 102*(21), 214106.
36. Zhang, X. C., Myers, E. B., Sader, J. E., & Roukes, M. L. (2013). Nanomechanical torsional resonators for frequency-shift infrared thermal sensing. *Nano Letters, 13*(4), 1528–1534.
37. Piller, M., Luhmann, N., Chien, M. -H., & Schmid, S. (2019). Nanoelectromechanical infrared detector. In *Optical Sensing, Imaging, and Photon Counting: From X-rays to THz 2019* (Vol. 11088, p. 1108802). International Society for Optics and Photonics.
38. Larsen, T., Schmid, S., Villanueva, L. G., & Boisen, A. (2013). Photothermal analysis of individual nanoparticulate samples using micromechanical resonators. *ACS Nano, 7*(7), 6188–6193.
39. Biswas, T. S., Miriyala, N., Doolin, C., Liu, X., Thundat, T., & Davis, J. P. (2014). Femtogram-scale photothermal spectroscopy of explosive molecules on nanostrings. *Analytical Chemistry, 86*, 11368–11372.
40. Schmid, S., Wu, K., Larsen, P. E., Rindzevicius, T., Boisen, A. (2014). Low-power photothermal probing of single plasmonic nanostructures with nanomechanical string resonators. *Nano Letters, 14*, 2318–2321.
41. Yamada, S., Schmid, S., Larsen, T., Hansen, O., & Boisen, A. (2013). Photothermal infrared spectroscopy of airborne samples with mechanical string resonators. *Analytical Chemistry, 85*(21), 10531–10535.
42. Chien, M. -H., Brameshuber, M., Rossboth, B. K., Schütz, G. J. & Schmid, S. (2018). Single-molecule optical absorption imaging by nanomechanical photothermal sensing. *Proceedings of the National Academy of Sciences, 115*(44), 11150–11155.
43. Jun, S. C., Huang, X. M. H., Manolidis, M., Zorman, C. A., Mehregany, M., & Hone, J. (2006). Electrothermal tuning of Al–SiC nanomechanical resonators. *Nanotechnology, 17*(5), 1506–1511.

44. Vig, J. R., Filler, R. L., & Kim, Y. (1997). Application of quartz microresonators to uncooled infrared imaging arrays. In: *Semiconductors and Semimetals* (Vol. 47, pp. 269–296). Elsevier.
45. Snell, N., Zhang, C., Mu, G., Bouchard, A., & St-Gelais, R. (2022). Heat transport in silicon nitride drum resonators and its influence on thermal fluctuation-induced frequency noise. *Physical Review Applied, 17*(4), 044019.
46. Piller, M., Sadeghi, P., West, R. G., Luhmann, N., Martini, P. Hansen, O., & Schmid, S. (2020). Thermal radiation dominated heat transfer in nanomechanical silicon nitride drum resonators. *Applied Physics Letters, 117*(3), 034101.
47. Rogalski, A. (2019). *Infrared and terahertz detectors.* Boca Raton: CRC Press.
48. Piller, M., Hiesberger, J., Wistrela, E., Martini, P., Luhmann, N., & Schmid, S. (2022). Thermal IR detection with nanoelectromechanical silicon nitride trampoline resonators. *IEEE Sensors Journal, 23*, 1–1.
49. Chien, M. -H. (2021). *Nanoelectromechanical photothermal microscopy and spectroscopy for single-molecule detection and imaging at room temperature.* PhD thesis, Technische Universität Wien.
50. Kurek, M., Carnoy, M., Larsen, P. E., Nielsen, L. H., Hansen, O., Rades, T., Schmid, S., & Boisen, A. (2017). Nanomechanical infrared spectroscopy with vibrating filters for pharmaceutical analysis. *Angewandte Chemie, 129*(14), 3959–3963.

Index

© The Author(s), under exclusive license to Springer Nature Switzerland AG 2023
S. Schmid et al., *Fundamentals of Nanomechanical Resonators*,
https://doi.org/10.1007/978-3-031-29628-4

Printed in the United States
by Baker & Taylor Publisher Services